# SpringerBriefs in Earth Sciences

More information about this series at http://www.springer.com/series/8897

Swapan Kumar Maity · Ramkrishna Maiti

# Sedimentation in the Rupnarayan River

Volume 1: Hydrodynamic Processes
Under a Tidal System

 Springer

Swapan Kumar Maity
Department of Geography
Nayagram P.R.M. Government College
Jhargram, West Bengal
India

Ramkrishna Maiti
Department of Geography
  and Environment Management
Vidyasagar University
Paschim Medinipur, West Bengal
India

ISSN 2191-5369          ISSN 2191-5377   (electronic)
SpringerBriefs in Earth Sciences
ISBN 978-3-319-62303-0          ISBN 978-3-319-62304-7   (eBook)
https://doi.org/10.1007/978-3-319-62304-7

Library of Congress Control Number: 2017945254

Printed on acid-free paper

This Springer imprint is published by Springer Nature
The registered company is Springer International Publishing AG
The registered company address is: Gewerbestrasse 11, 6330 Cham, Switzerland

# Preface

A river is a dominant geomorphic process and is engaged in reshaping the earth's surface by transfer of mass and energy for its long geologic past. Sedimentation on river bed is very important to control the geomorphological, hydrological and ecological characteristics of the river and the surrounding areas. At present, river sedimentation is one important problem in the world and has been accelerated by human activities like, forest cutting, unscientific agriculture, mining, road construction and rapid urbanization. Sedimentation creates a variety of detrimental problems and impacts on society, economies and the environment. The main aim of the book is to understand and explain the causes, mechanisms and extent of river sedimentation in connection to all the hydrodynamic characteristics of the river.

Channel forms and patterns including symmetrical and asymmetrical shape of the cross-sections, channel diversion and flow separation, distribution of depth, width-depth ratio etc. have been studied and measured in detail. Hydraulic characteristics of the stream like, nature and pattern of flow, distribution of water velocity, seasonal variation of water velocity and discharge have been measured carefully directly in the field. Tidal data have been measured and collected at different gauge stations to understand the nature and characteristics of tidal range, tidal prism and tidal asymmetry. Suspended and bed load of the river are measured and estimated following suitable techniques. Sediment grain size has been calculated by collecting sediment samples and analyzing them using sieving technique.

Sedimentation in the tidal river is the result of combined interaction of riverine and marine processes. Channel asymmetry, channel diversification and separation of flow, seasonal variation of water discharge during high and low tide, tidal asymmetry and associated variation of sediment transporting capacity are the main causative and controlling factors of sedimentation. The result of the work will be extremely beneficial to the engineers, hydrologists, planners and other concerned authorities, working on the aspects of sedimentation and management of associated problems in different climatic, geological, geomorphological, hydrological, bathymetric and hydraulic characteristics.

Midnapur, West Bengal, India                                        Swapan Kumar Maity
                                                                              Ramkrishna Maiti

# Acknowledgements

We would like to express our humble regards and thankfulness to some of our students Sanjay, Arindam, Pratap, Subrata, Tarun, Subhasis and Suman for their continuous help and support during the entire research work. Without their untired help and active participation, the completion of our research work could have been impossible. We also like to extend our immense sense of gratitude to Dr. Animesh Majee, Dr. Prasenjit Bhunia, Mr. Sarbeswar Haldar, Mr. Rajkumar Ghorai, Mr. Alakesh Samanta and Sk. Nurul Alam for their assistance during conducting the research, technical support and valuable suggestions.

We convey our appreciation and thanks to Dr. Moumita Moitra Maiti, Assistant Professor, Department of Geography, Raja N.L. Khan Women's College, for her immense support throughout the work.

We are highly obliged to the concerned authority of Kolaghat Thermal Power Station, Kolkata Port Trust, Sub-divisional Office at Tamluk and Ghatal, Geological Survey of India, Survey of India for providing us necessary documents and data. We are especially thankful to the Central Research Facility of Indian Institute of Technology (Kharagpur) for helping us in sedimentological analysis.

We convey a lot of thanks to all the members of our family for their encouragement, co-operation and continuous moral and emotional supports.

Midnapur, West Bengal, India                                         Swapan Kumar Maity
                                                                     Ramkrishna Maiti

# Contents

# Abbreviations

| | |
|---|---|
| ASTM | American Society for Testing Materials |
| CRF | Central Research Facility |
| DPM | District Planning Maps |
| GPS | Global Positioning System |
| GSI | Geological Survey of India |
| IIT | Indian Institute of Technology |
| IRS | Indian Remote Sensing (Satellite) |
| KPT | Kolkata Port Trust |
| KTPP | Kolaghat Thermal Power Project |
| KTPS | Kolaghat Thermal Power Station |
| LDA | Linear Discriminate Analysis |
| LISS | Linear Imaging Self-Scanning Sensor |
| NATMO | National Atlas and Thematic Mapping Organization |
| NRSA | National Remote Sensing Agency |
| PCA | Principal Component Analysis |
| SI | Sinuosity Index |
| SOI | Survey of India |
| WBPDCL | West Bengal Power Development Corporation Limited |

## Symbols

| | |
|---|---|
| $D$ | Median grain size (m) |
| $D$ | Water depth |
| $H$ | Tidal wave amplitude |
| $H$ | Constant hydraulic head |
| $P$ | Tidal prism |
| $Q$ | Water discharge |
| $R$ | Hydraulic radius |
| $S$ | River bed slope |
| $V$ | Mean velocity |

| W | Width |
|---|---|
| g | Gravitational acceleration |
| h | Horizontal distance |
| $h$ | Depth of flow |
| r | Reduced level |
| $r$ | Co-relation co-efficient |
| $\theta$ | Slope |
| $\phi$ | Sediment grain size unit |
| $\rho$ | Water density |
| $\rho s$ | Sediment density |
| $\mu$ | Dynamic viscosity |
| $\gamma$ | Specific weight of the water |
| $\gamma_s$ | Specific weight of sediment |
| $\Omega$ | Stream power |
| Fr | Froude number |
| Re | Reynolds number |
| $Q_0$ | Additional discharge during low tide in rainy season |
| $(Q + Q_0)^m$ | Sediment transport power during rainy season low tide |
| $Q^m$ | Sediment transport power during flood tide in dry season |
| $Q_S$ | Suspended sediment transport |
| $Q_T$ | Bed load transport |
| $f^1$ | Friction factor |

# List of Figures

# List of Tables

# Chapter 1
# Introduction

**Abstract** River *sedimentation* is one of the major water related problem in the world and also in India. Most of the rivers in India are facing the problems of rapid sedimentation and associated socio-environmental hazards. In southern part of West Bengal a large number of rivers like, Damodar, Rupnarayan, Kangsabati and Haldi are being sedimented rapidly and facing the problems of drainage congestion and drainage decay, deterioration of navigability, unavailability of water resources, water storage and resultant flood, river bank erosion, loss of settlements and properties and social dislocation. These *problems of sedimentation* are more serious and alarming in *Rupnarayan River*, mainly at the lower reach, from Kolaghat to Geonkhali. During the last 26 years, the area has experienced 28.71 million m$^3$ shoaling causing the deterioration and incapacitation of the river. The total shoaled up area has increased from 15.41 km$^2$ to 57.35 km$^2$ between the years 1973 and 2016. *Causes, mechanisms* and *extent* of sedimentation in the lower reach are explained and understood with the detailed study of channel forms and patterns, stream hydraulics, tidal character, sediment load, sediment grain size and related critical and available shear stress and identification of the environment of sediment deposition.

**Keywords** Problems of sedimentation · Causes and mechanisms · Lower reach · Rupnarayan river

## 1.1 Prologue

*Stream channels* are very dynamic features of the landscape, changing their shape, size, planform, morphology and bed forming material with time and in accordance with the changes of water flow and *sediment load*. The mobile materials that make up the bed of the stream, banks and floodplain has been transported and deposited there by the stream and can again be moved in appropriate conditions. In a relatively stable stream channel the volume of water flow and *sediment flux* always maintains a balanced condition over time (McManus 1985). Any type of changes of

© The Author(s) 2018
S. Kumar Maiti and R. Maiti, *Sedimentation in the Rupnarayan River*,
SpringerBriefs in Earth Sciences, https://doi.org/10.1007/978-3-319-62304-7_1

either of these two factors, may lead to the channel adjustment with depth and width, slope, meander pattern and bed composition consequently. The dimension and rate of these adjustments are mostly governed by the extent and rate of change in the volume of water flow and sediment load, resistance of the bank and bed form and threshold of change within the concerned fluvial system. An ongoing and most significant problem in the management of freshwater ecosystems is *sedimentation*. Sedimentation within the river bed and catchments system is very important and challenging as it is a crucial parameter to determine the water system dynamics (Salas and Shin 1999). Both natural and man induced factors control the rate and character of accumulation of sediments, whereas the river bed morphology determines the prevalence of *sediment erosion*, transportation and zones of accumulation (Tylmann 2004). If the supply of sediment from the upstream area is same to the volume of sediment which is being entrained, then the sediment coming in to a particular river segment will replace the entrained sediment (Lane 1955; Charlton 2007; Clayton and Pitlick 2008; Mayoral 2011). But in case of huge volume of sediment discharge from upstream area, the river response by depositing the sediments in order to increase the steepness of the gradient and its water velocity so that the river becomes more competent (Smith 1974; Mackin 1948). *Critical shear stress* is very significant indicator of *entrainment of sediment*, which could potentially result in the channel bed or banks erosion (Buffington and Montgomery 1997; Ritter et al. 2002; Church 2006; Charlton 2007; Mayoral 2011). The rate of sediment transportation and deposition in *estuarine environments* are greatly affected by the tidal currents and volume of river discharges (Hall et al. 1987; Mayoral 2011). The mechanisms of transportation and deposition of sediment particles are influenced by different parameters including roundness, sphericity, surface texture, heavy mineral ratio and presence of biogenic components (Folk and Ward 1957; Friedman 1979; Martins 1965).

## 1.2   Sedimentation at the Lower Reach of the Rupnarayan River

*Sedimentation* in river is one of the important water related problem in the world and it is being accelerated by different *anthropogenic activities* such as forest cutting, unscientific agricultural practices, mining activities, construction of road and rapid development of urban and suburban areas. In Indian territory there are 12 large, 46 medium and numerous minor river basins which have an important contribution for the distribution of surface water and their utilization for different purposes along with the storage of ground water potential (Venkateswara Rao et al. 2014). The major problem of each and every river basin is rapid rate of *sedimentation* and associated socio-environmental hazards, thus proper understanding and scientific techniques are required for the preservation of these river basins to meet the present and future needs.

Due to typical topographical position, West Bengal has a dense web of fluvial arteries being criss crossed by myriads of channels, streams, rivulets, torrents somewhere becoming almost a labyrinth. In *southern part of West Bengal* most of the rivers like, *Damodar, Dwarakeswar, Silabati, Rupnarayan, Kangsabati* and *Haldi* are being sedimented rapidly and facing the problems of *drainage contestation and drainage decay, deterioration of navigability, inadequate availability of water resources to meet agricultural need, water storage and resultant flood* etc (Mukhopadhyay and Dasgupta 2010). Problems of sedimentation are more severe and alarming in *Rupnarayan River*, mainly in the *lower reach of the river*. River Silabati and Dwarakeswar meet near Bandar (Ghatal) and the combined flow is named as Rupnarayan (Fig. 1.1), which joins river Hoogly at Geonkhali covering a distance of 78 km with a catchment area of 10930 km$^2$.

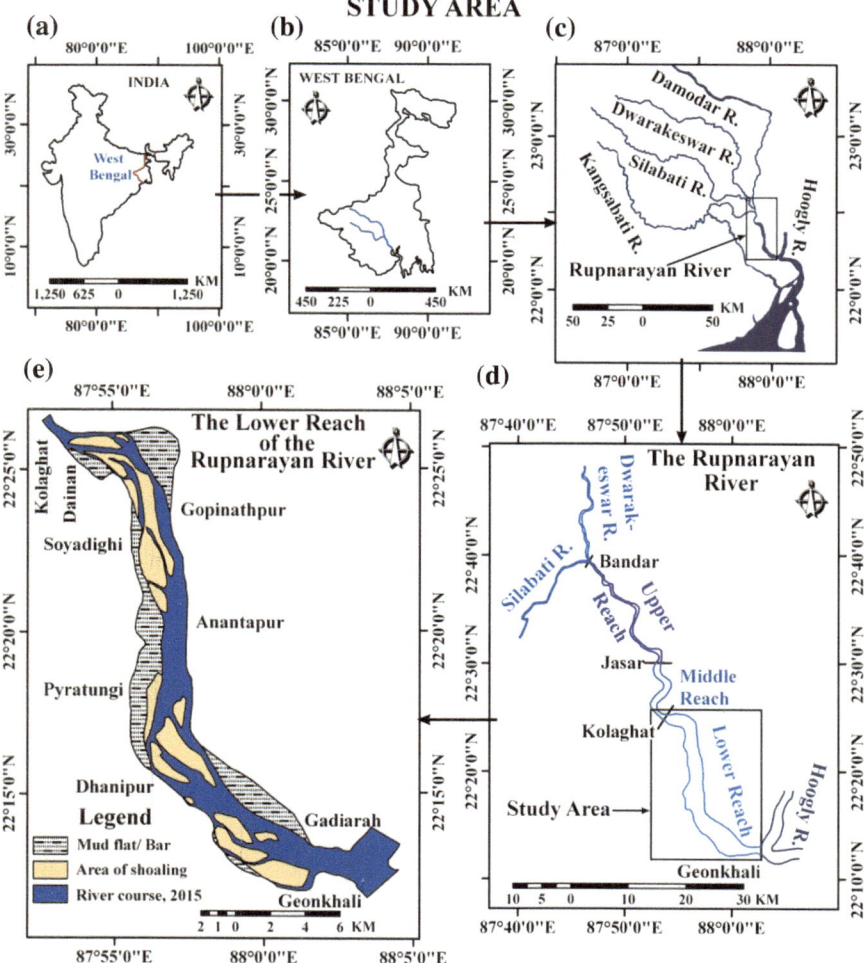

**Fig. 1.1** The study area

The entire Rupnarayan River has been divided into three reaches based on the configuration and geometry of the channel and other properties of the river (Fig. 1.1). The upper reach is 28 km long and is extended between Bandar and Jasar whereas the middle reach extends between Jasar and Kolaghat (Dainan) with a length of 10 km. The lower reach, the present study area extends from *Kolaghat (Dainan) to Geonkhali* with a length of 40 km and is bounded by $22^0 12'$N to $22^0 26'$N and $87^0 53'$E to $88^0 03'$E (Figs. 1.1 and 1.2). It is further divided into five sub-reaches (Geonkhali to Dhanipur, Dhanipur to Pyratungi, Pyratungi to Anantapur, Anantapur to Soyadighi and Soyadighi to Dainan reach) based on the necessity of the study.

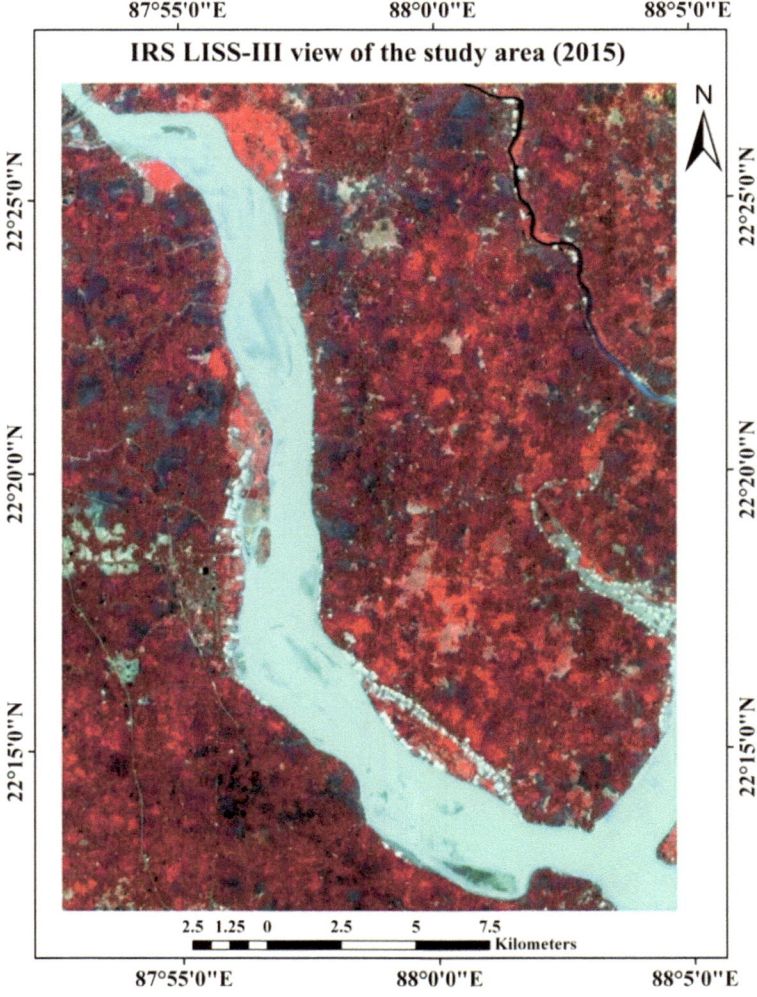

**Fig. 1.2** IRS LISS-III view of the study area (February, 2015). (*Source* National Remote Sensing Agency, India)

**Table 1.1**  Total shoaling and scouring area in different sub-reaches (1990–2016)

| Sub-reaches | Total shoaling (million m³) | Total scouring (million m³) | Net shoaling (million m³) |
|---|---|---|---|
| Geonkhali to Dhanipur | 12.26 | 3.75 | 8.51 |
| Dhanipur to Pyratungi | 10.85 | 3.06 | 7.79 |
| Pyratungi to Anantapur | 6.18 | 2.74 | 3.44 |
| Anantapur to Soyadighi | 5.12 | 1.96 | 3.16 |
| Soyadighi to Dainan | 8.73 | 2.92 | 5.81 |
| Total (40 km) | 43.14 | 14.43 | 28.71 |

*Source* K.P.T, K.T.P.P. and field survey

In the recent past, the lower reach of the river has been deteriorated and has lost its capacity due to continuous sedimentation and development of *shoal area*. During the last 26 years (1990–2016) the area has experienced a net *sediment deposition* of *28.71 million m³* with 43.14 million m³ shoaling and 14.43 million m³ scouring (Table 1.1). The data shown in Table 1.1 reveal that Geonkhali to Dhanipur, Dhanipur to Pyratungi and Soyadighi to Dainan sub-reaches are mostly affected by shoaling than that of scouring. In all other sub-reaches (Pyratungi to Anantapur and Anantapur to Soyadighi) the amount of shoaling is comparatively less (Table 1.1).

The rate of sedimentation varies widely in different seasons at different places in the lower reach. Recent measurement (in the year 2016) indicates that the sedimentation rate is high near Dhanipur compared to other places in all the seasons. Near Anantapur region the rate of sedimentation is comparatively low. Again, near Soyadighi and Kolaghat region the sedimentation rate has increased to some extent (Table 1.2). Sedimentation rate is high in dry season (pre-monsoon and post-monsoon) compared to monsoon season. The rate of sedimentation varies between 2.38 mm/h to 3.14 mm/h in pre-monsoon, 1.31 mm/h to 3.0 mm/h in monsoon and 2.39 mm/h to 3.56 mm/h in post-monsoon season (Table 1.2). Highest and lowest sedimentation rate (3.56 mm/h and 1.31 mm/h respectively) is measured during post-monsoon and monsoon season near Dhanipur and Anantapur respectively (Table 1.2). The deterioration is likely to hamper the easy discharge of

**Table 1.2**  Spatial and seasonal variation of sedimentation rate (2016)

| Sedimentation rate (mm/hour) | | | | | | |
|---|---|---|---|---|---|---|
| Reach | Geonkhali | Dhanipur | Pyratungi | Anantapur | Soyadighi | Dainan/Kolaghat |
| Pre-monsoon | 2.86 | 3.14 | 2.96 | 2.38 | 2.64 | 2.38 |
| Monsoon | 2.16 | 3.0 | 1.94 | 1.31 | 1.56 | 1.83 |
| Post-monsoon | 3.07 | 3.56 | 2.87 | 2.39 | 2.68 | 3.20 |

*Source* Field survey

fresh water brought down by Dwarakeswar, Silabati, Kangsabati and other rivers and is also likely to affect the *Hoogly River* adversely to some extent. Problems of sedimentation in this region were studied by Sinha and Basu (1960) in a mathematical model by Harmonic Method which helps one to get an insight into the *tidal mechanism*. The Calcutta Port Commissioners conducted the hydrodynamic survey of the lower reach of the river in 1968. The Irrigation and Waterways Department of the government of West Bengal conducted the survey for the entire river in 1962–1963 and repeated the same for portions of the upper reach of the river in 1970 and constructed a mathematical model for the entire *Rupnarayan River* with the cross-sectional survey data. In the recent past an intensive survey was conducted jointly by Kolkata Port Trust (KPT) and Kolaghat Thermal Power Project (KTPP) in this area.

### 1.2.1  Physical Background of the Catchment Area of the Rupnarayan River

The catchment area of the Rupnarayan River experiences typical tropical monsoonal climate characterized by annual average rainfall of 1320–1630 mm and the annual temperature ranges from 7 to 45 °C. The upper catchment of the river is characterized by steep gradient to moderately steep gradient, but in the lower catchment, the gradient is almost gentle to flat (<10 m per km). Most of the portions of the lower catchment are below the elevation of 10 m (O'Malley 1995). *Granite and gneiss* along with shale and sandstone of lower Gondwana, micaschist, phyllite and quartzite are the important constituent rocks of the upper catchment area of the river basin. Middle portion of the catchment is mainly formed of granite and gneiss but phyllite schist, sandstone, shale, grit, micaschist, amphibolites, newer dolerite and conglomerate are also important rock components (Fig. 1.3). Laterite, pleistocene sediments and older alluvium are the main constituents of the lower middle portion of the region. The lower portion (mainly the eastern and southern portion) is composed of *newer alluvium* (O'Malley 1995; Mukhopadhyay and Dasgupta 2010) (Fig. 1.3).

Geomorphologically, the upper part of the catchment is composed of dissected hill slopes, highly gullied land, inter hill valley and some scattered residual hills, hillocks and mounds with pediments (Fig. 1.4). Pediplain is the main geomorphic unit in the middle catchment and highly gullied lands are found in scattered form. Resudial hills, hillocks and mounds are common geomorphic features in this region. The lower catchment of the river is composed of lower alluvial plain and deltaic flood plain (O'Malley 1995; Mukhopadhyay and Dasgupta 2010) (Fig. 1.4). Most of the region of the upper and middle catchment of the Rupnarayan River is constituted of reddish lateritic soil. The coarse lateritic soil and clay laterites are predominant type in the west and north-west region. Younger and newer alluvial soils are found mostly in the south and south-east region of the catchment area, which are enriched by silt and clay deposition (O'Malley 1995).

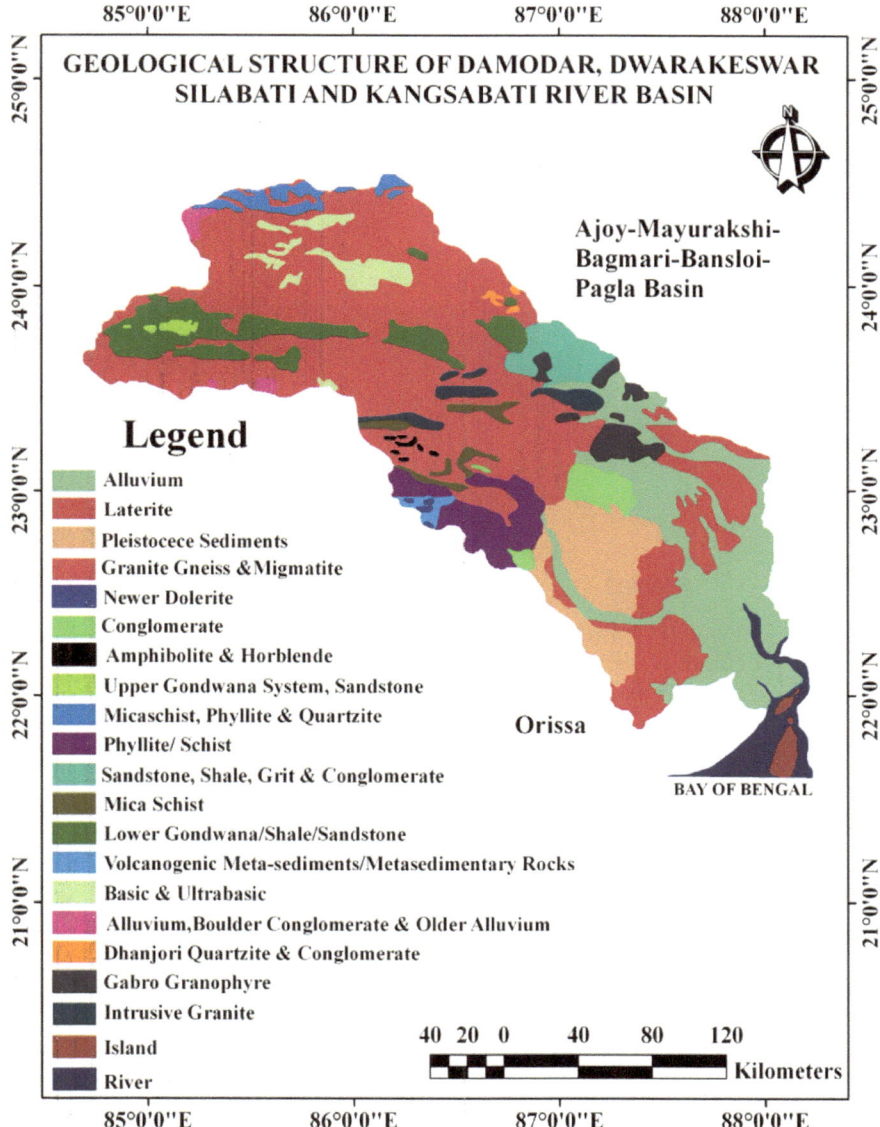

**Fig. 1.3** Geological structure of the catchment area (Maity and Maiti 2016). (*Source* Geological Survey of India, Kolkata)

The river Rupnarayan receives water from several other rivers. Among them, Dwarakeswar, Silabati, Damodar and Kangsabati are important (Fig. 1.5). Dudhbhariya, Dagra, Gandheshwari, Haringmuri, Chottokandar, Kana and Mundeshwari (main bifurcated channel of river Damodar) are the main tributaries of Dwarakeswar River. The river Purandar, Jaipanda, Amudar, Ketia khal, Sundara,

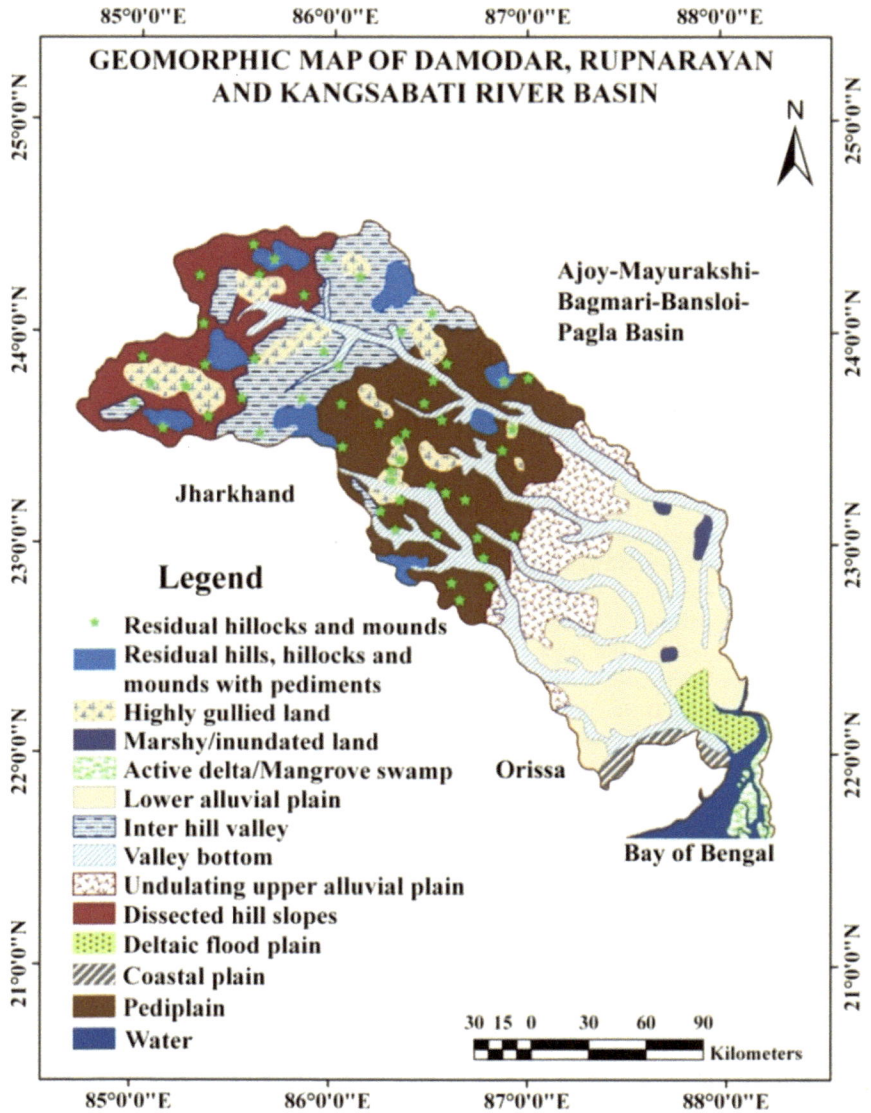

**Fig. 1.4** Geomorphological map of the catchment area of Rupnarayan River (Maity and Maiti 2016). (*Source* Geological Survey of India, Kolkata, 1991)

Parang and Kubai are the major tributaries of the river Silabati. A left bifurcated channel of Kangsabati River meets the River Rupnarayan few km downstream from Bandar. So, huge volume of water and sediment of all these rivers is discharged to the River Hoogly through the Rupnarayan River (O'Malley 1995; Mukhopadhyay and Dasgupta 2010) (Fig. 1.5).

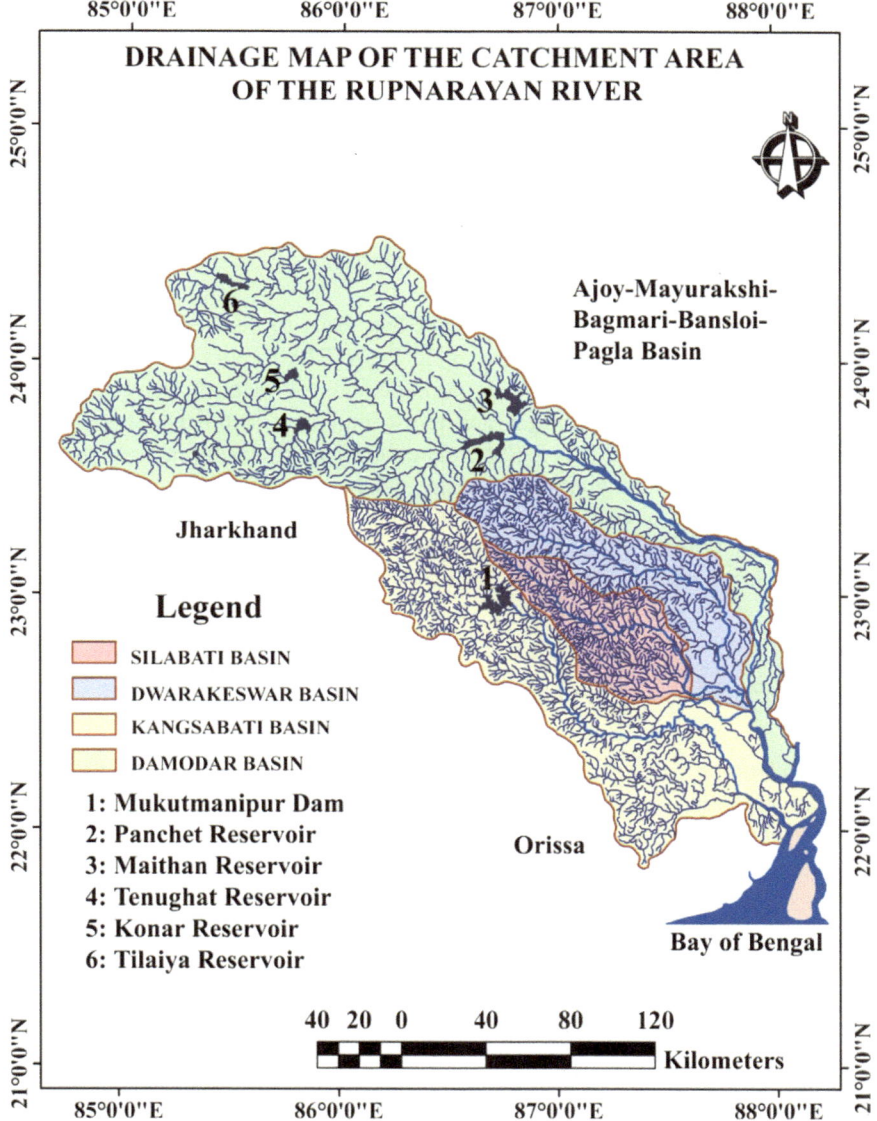

**Fig. 1.5**  Drainage map of the catchment area of Rupnarayan River. (*Source* SRTM DEM)

The river is not so big in the upper reach (maximum width is 280 m) but it gradually expands on its downward journey (maximum width is 4250 m) (Mukhopadhyay and Dasgupta 2010) (Fig. 1.1). The river is influenced by the semi-diurnal tide throughout its course from Bandar to Geonkhali (78 km). During the dry months brackish water is found as far as Kolaghat, but during rainy months the salt water is driven out by the volume of fresh water brought down from upper catchment. The river is navigable by boats and small steamers, throughout the year. Several islands are found in the river channel, where accretions in the shape of grass-covered chars are frequent, especially near Soyadighi, 9 km north of Tamluk. Throughout this stretch the river is interspaced with inter-tidal flats which keep on shifting, decaying, growing, and contributing to the continuous changes of the river morphology (Sinha and Basu 1960). Generally the channel regime is being continuously moulded under the interaction of the tide and upland discharge and the water way has been subjected to the influences of numbers of engineering structures, like four bridges built over piers, cluster of piles built for power transmission lines (Maity and Maiti 2017).

Continuous sedimentation on the river bed leads to the development of shoal area at the lower reach of the Rupnarayan River. The location and areal extent of the shoal is changing to adjust with the fluctuation of water discharge and stream energy, changes of river gradient and redistribution of sediments. Topographical maps (Survey of India), IRS LISS-III satellite images have been used to identify and understand the changing nature of the shoal areas. The shifting and changes of the areal extent of the shoals is shown in Table 1.3 and Fig. 1.6. The course of the river and the thalweg position (deepest portions) is also changing continuously to adjust with the development and shifting of shoal area.

## 1.2.2   Problems of Sedimentation in the Lower Reach

Sedimentation creates different types of effects and problems (shortage of water during neap tide, hindrance of easy discharge of water and *upstream flood*, *navigation difficulties*, *river bank erosion* and *loss of settlements* etc.) on environment and society.

### 1.2.2.1   Water Shortage to Kolaghat Thermal Power Station (KTPS)

The entire amount of make-up water for Kolaghat Thermal Power Station is drawn from *Rupnarayan River* through Intake Pump House. The withdrawal of water by West Bengal Power Development Corporation Limited (WBPDCL) from this river is 1,50,000 $m^3$/day (61.285 Cusec) and can only be possible during high tide for a period of 3 to 4 h. Due to rapid sedimentation and formation of huge shoal over the years covering major portion of the river around the Dolphin Mouth of the Pump House, acute problem is being faced for getting adequate water for the plant (Fig. 1.7a). Thus, the generation of power has suffered due to shortage of water, especially during lean season neap tides (March to May). In the month of March, 2004 for about 2 days raw water feeding for power generation remained cut-off causing great loss of generation. Again total obstruction to the flow of water into Dolphin mouth happened on 9th, 10th and 11th September 2004, owing to reduction of water level during the neap tide. The situation was so alarming that it leads to shut

**Table 1.3** Shifting and changes of the areal expansion of the Shoal Area (Maity and Maiti 2017)

| Year | Area of shoal (km²) | | | | | | | | | | | |
|---|---|---|---|---|---|---|---|---|---|---|---|---|
| | S-1 | S-2 | S-3 | S-4 | S-5 | S-6 | S-7 | S-8 | S-9 | S-10 | S-11 | S-12 |
| 1973 | 0.88 | 0.02 | 3.08 | 0.13 | 0.46 | 0.18 | 1.09 | 0.8 | 0.17 | 0.5 | 0.14 | 2.0 |
| 1990 | 1.11 | 0.18 | 0.77 | 3.64 | 1.43 | 1.39 | 0.25 | 0.18 | 2.47 | 1.64 | 0.38 | 0.55 |
| 2000 | 1.1 | 1.06 | 0.18 | 7.14 | 0.24 | 1.48 | 11.32 | 1.51 | 3.15 | 1.78 | 6.83 | 0.82 |
| 2016 | 2.47 | 1.68 | 0.19 | 8.13 | 6.82 | 0.82 | 14.26 | 1.35 | 2.0 | 1.32 | 4.47 | 1.96 |

| Year | Area of shoal (km²) | | | | | | | | | | Total |
|---|---|---|---|---|---|---|---|---|---|---|---|
| | S-13 | S-14 | S-15 | S-16 | S-17 | S-18 | S-19 | S-20 | S-21 | S-22 | |
| 1973 | 0.58 | 4.33 | 0.61 | 0.41 | – | – | – | – | – | – | 15.41 |
| 1990 | 2.21 | 1.98 | 4.1 | 0.38 | 4.78 | 1.88 | 1.98 | 7.75 | 3.85 | 0.56 | 43.43 |
| 2000 | 7.15 | 0.98 | 0.72 | 0.7 | – | – | – | – | – | – | 46.15 |
| 2016 | 8.17 | 0.16 | 1.86 | 0.89 | 2.81 | 1.02 | 4.53 | 0.56 | – | – | 57.35 |

*Source* SOI topographical maps, IRS LISS-III image and field survey

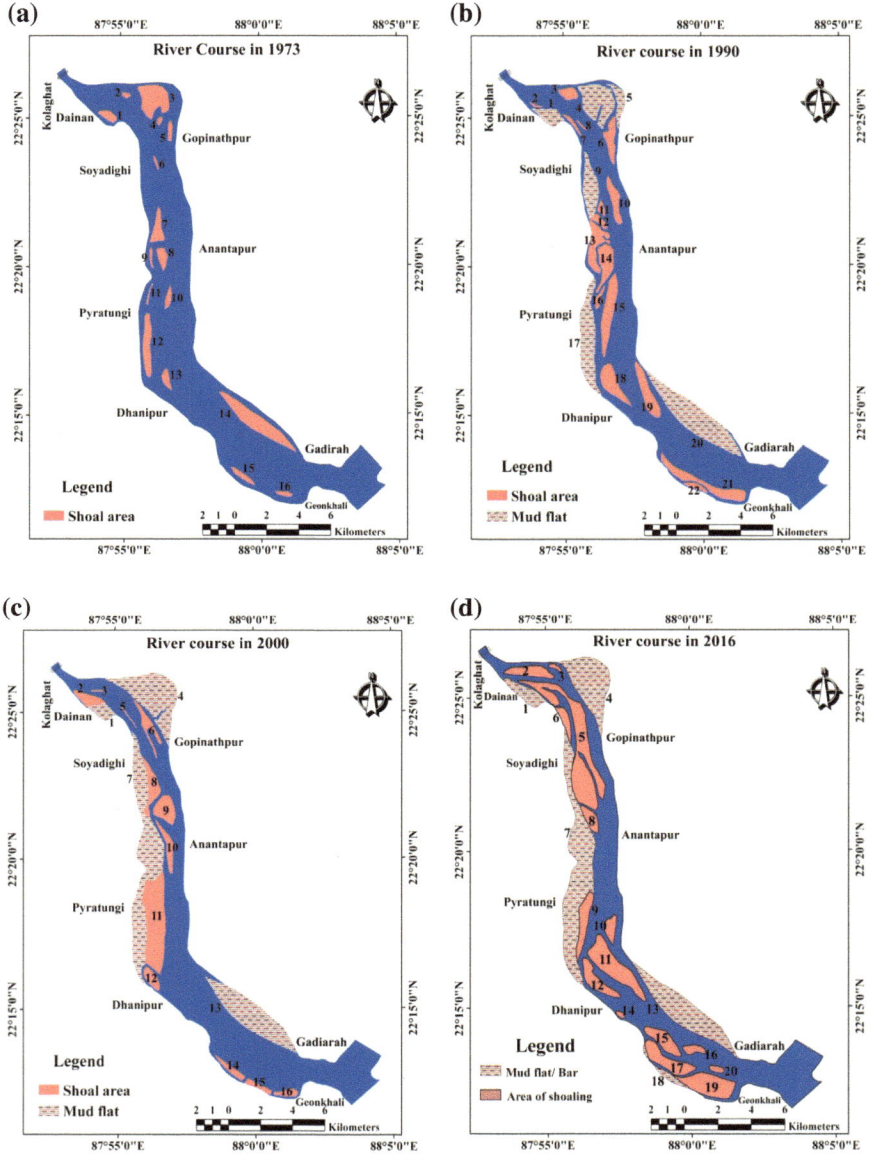

**Fig. 1.6** Shifting of shoal area and changes of its areal expansion (Maity and Maiti 2017), **a** Shoal area in 1973 (*Source* SOI Topographical maps), **b** Shoal area in 1990 (*Source* IRS LISS-III image), **c** Shoal area in 2000 (*Source* IRS LISS-III image), **d** Shoal area in 2016 (*Source* IRS LISS-III image and field survey)

down the generation of 3 Units. Under these circumstances, the Authority of KTPS had to engage dredgers for dredging a channel from Dolphin mouth to main stream up to dependable depth for getting required water during the ebb tide.

#### 1.2.2.2   Hindrance in Easy Discharge and Resultant Flood

Continuous sedimentation and development of shoal area has raised the bed level of lower reach of the Rupnarayan River, which has an adverse impact on the free discharge of upstream tributaries (Silabati, Dwarakeswar, Kangsabati and Damodar) causing water storage and increasing *flood possibility* in the lower catchment of these tributaries. Ghatal, Daspur-I, Daspur-II, Chandrakona-I, Chandrakona-II and Keshpur blocks of Paschim Medinipur district are badly affected by the flood of Silabati River (Fig. 1.7b). Daspur-II block is also affected by the flood of Kangsabati River. Arambag of Hoogly district is another flood affected region. The intensity of flood of these regions increases due to sedimentation and hindrance in free discharge through lower reach of the Rupnarayan River. So, a *water logging* condition in the lower catchment of these upstream tributaries arises due to hindrance of water flow mainly during rainy season when discharge from land is enormous.

#### 1.2.2.3   Navigation Difficulties

The sedimentation in water courses can also make them unsuitable for navigation without regular dredging work. The lower reach of the Rupnarayan River has been incapacitated and facing the problem of *navigation difficulties* due to continuous sedimentation and decrease of river depth. Vessels and other modes of water transport can't move freely on the river because of lack of sufficient water depth. At some places, the problem is so acute that the vessels have to stay far away from the river bank and the people have to walk a long distance (more than 1 km) on the river bed to reach the river bank (Fig. 1.8a). Though *dredging* is done to maintain the adequate depth of the river, but it is infrequent because it is costly to operate.

**Fig. 1.7** Shortage of water near K.T.P.P (**a**) and problem of flood (**b**)

**Fig. 1.8** Navigation difficulties (**a**) and loss of settlement due to bank erosion (**b**)

### 1.2.2.4   River Bank Erosion, Bank Failure and Loss of Settlement and Properties

*Erosion of the river bank*, *bank failure* and *destruction of settlements* is important problems related to the sedimentation on river bed and incapacitation of the river. When sediment is deposited and shoal area is developed near one bank of the river, then the main course and thalweg position of the river is shifted towards the other bank. The shifting of the main channel and amassing of water increase the pressure on one bank, leading to bank failure and *erosion of the bank*. This problem becomes more dangerous and unavoidable during monsoon season when huge terrestrial discharge increases the volume of river water and creates immense pressure on the river bank. Recently, a huge bank failure occurred near Kolaghat (Fig. 1.8b) and Tamluk region, causing great loss of settlements and properties. Near Kolaghat region, the river has taken a sharp bend and erosion concentrates towards the concave right bank. During the time of discharge of huge volume of water immense pressure was created on the right bank and the bank failure occurred.

## 1.3   Objectives of the Present Work

The hydraulic and geometric characteristics of the region are complex, chaotic, unpredictable and quite different from the normal stream flow due to tidal influence. The basic objective of the present study is to understand the *causes, mechanisms and extent of sedimentation* at lower reach of the Rupnarayan River. For the fulfilment of the basic objectives, the present work is spread into the following:

1. To understand the hydraulic and geometric characteristics of the river and the nature of interaction of *fluvial and marine processes (tidal process)* in connection to spatial and seasonal variation of stream energy (shear stress), sediment load, sediment texture and sedimentation rate.

2. Studying the *environment of sediment deposition* and the hydrodynamic processes operating during deposition by understanding sources of sediments.


## 1.4   Applied Methodology

Different methods and techniques have been adopted for the fulfillment of the basic
objectives of the present study. *Twenty five (25) cross-section* lines, at an interval of
1.5 km, were fixed along the lower reach of the Rupnarayan River, based on
meandering pattern; different geomorphic units; area of shoaling and scouring and
ease of data collection etc. Intensive survey and monitoring was made at an interval of
20 m along these lines throughout the year and required data (depth, water velocity
etc) were generated using various survey instruments. Six (6) stations (Kolaghat,
Soyadighi, Anantapur, Pyratungi, Dhanipur and Geonkhali) were taken to be fixed
along the lower reach of the Rupnarayan River and sixty (60) sites, based on grain size
variation; color variation; geomorphic unit and area of shoaling and scouring were
selected and *one hundred eighty (180) sediment samples* were collected in three
seasons (Pre-monsoon, Monsoon and Post-monsoon) for sedimentological analysis.
All the required parameters (sedimentation rate, water discharge, tidal water level,
suspended sediment load and transport, bed load transport, critical and available shear
stress etc) were monitored and estimated at these six (6) stations in three seasons.

1. Depth of river was measured by *Echo Sounder* interfaced with position fixing
system. Dumpy level was used for the collection of elevation related data.
2. Water velocity was measured by *floating method* and using *Digital current
meter*.
3. Tidal water level was measured at different gauge stations at an interval of one
hour.
4. Water samples during high and low tide have been collected to know the
concentration of *suspended load* in water.
5. *Rate of sedimentation* in three seasons was measured by fixing the wooden tray
on river bed for few days.
6. Bed sediment samples have been collected for grain size analysis by sieving
technique and to know the mineralogical composition of sediments by X-Ray
Diffraction technique.

The analysis and interpretation of the collected and measured data are done in
laboratory using well accepted formulae and equations of different researchers in
order to synthesize all the studied parameters in a holistic manner for understanding
the process operating in the study area.

## 1.5   Data Used in the Study

Necessary data have been collected from the following sources-

a) A good amount of *primary data* have been generated in the field using various survey instruments, Echo sounder, Digital current meter, floating method, GPS receiver etc.

b) A large amount of *primary data* was generated by analysis of collected water and sediment samples in laboratory.

c) *IRS LISS-III satellite imageries* were procured from National Remote Sensing Agency (NRSA), Hyderabad, India for the preparation of various maps.

d) *Topographical maps* (73N/5, 73N/6, 73N/7, 73N/8, 73N/9, 73N/10, 73N/11, 73N/12, 73N/13, 73N/14, 73N/15, 73N/16, 79B/4 and 73C/1) were collected from *Survey of India, Kolkata.*

e) *District Planning Maps (DPM)* have been taken from National Atlas and Thematic Mapping Organization (NATMO).

f) Soil and Land use Maps were available from National Bureau of Soil Survey and Land Use planning.

g) *Geological maps* are collected from Geological Survey of India (GSI), Kolkata.

h) *Tidal gauge height* and *water discharge* data were procured from the Office at Tamluk and Ghatal, under the Irrigation and Waterways Department of Government of West Bengal.

i) *Bathymetric, tidal, sediment and discharge data* have been collected from the concerned authority of Kolaghat Thermal Power Project and Kolkata Port Trust.

## References

Buffington J, Montgomery D (1997) A systematic analysis of eight decades of incipient motion studies, with special reference to gravel—bedded rivers. Water Resour Res 33(8):1993–2029

Charlton R (2007) Fundamentals of fluvial geomorphology Routledge, New York, p 234

Church M (2006) Bed material transport and the morphology of alluvial river channels. Annu Rev Earth Planet Sci 34:325–354

Clayton J, Pitlick J (2008) Persistence of the surface texture of a gravel—bed river during a large flood. Earth Surf Process Landf 33(5):661–673. doi:10.1002/esp.1567

Folk RL, Ward MC (1957) Brazos River bar (Texas): a study in the significance of grain size parameters. J Sediment Petrol 27(1):3–27

Friedman GM (1979) Differences in size distributions of populations of particles among sands of various origins. Sedimentology 26:3–32

Hall MJ, Nadeau JE, Nicilich M (1987) The use of trace metal content to verify sediment transport from Delaware Bay on to the New Jersey inner shelf. J Coast Res 3:469–474

Lane EW (1955) Design of stable alluvial channels, transactions. Am Soc Civil Eng 120 (2776):1234–1260

Mackin JH (1948) Concept of the graded river. Geol Soc Amer Bull 59:463–512

Maity SK, Maiti RK (2016) Understanding the Sources of sediments from mineral composition at lower reach of the Rupnarayan River, West Bengal, India– An X-Ray Diffraction (XRD) based analysis. GeoResJ 9(12):91–103

Maity SK, Maiti RK (2017) Sedimentation under variable shear stress at lower reach of the Rupnarayan River, West Bengal, India. Water Sci 31:67–92

Martins LR (1965) Significance of skewness and kurtosis in environmental interpretation. J Sediment Petrol 35(3):768–770

Mayoral H (2011) Particle size, and Benthic invertebrate distribution and abundance in a gravel-bed River of the Southern Appalachians. Geosciences Theses. Paper 31

McManus J (1985) Grain size determination and interpretation. In: Tucker M (ed) Techniques in sedimentology. Blackwell Scientific Publication, London, pp 63–85

Mukhopadhyay SC, Dasgupta A (2010) River dynamics of West Bengal, phys asp Vol.1, pp 1–220

O'Malley LSS (1995) Physical aspects. Midnapore, Published by West Bengal District Gazetteers, Bengal District Gazetteers, pp 1–141

Ritter D, Kochel R, Miller J (2002) Process geomorphology. McGraw-Hill Companies, New York, p 560

Salas JD, Shin H (1999) Uncertainty analysis of reservoir sedimentation. J Hydraul Eng 125: 339–350

Sinha G, Basu AN (1960) A study of the tidal propagation through the Rupnarayan river with the help of Mathematical Model, B.B.I. & P. pub. no110, Proc.41$^{st}$ Annual Research Session, Vol IB (Hydraulics)

Smith KG (1974) Erosional processes and landforms in Dadland National Monument, South Dakota. Geol Soc Amer Bull 69:975–1008

Tylmann W (2004) Estimating Recent Sedimentation Rates using Pb—210 on the example of Morphologically Complex Lake. Geochronometria 23:21–26

Venkateswara Rao S, Sastry PG, Ghorpade VG (2014) Reservoir sedimentation and concerns of stakeholders. Res J Eng Sci 3(2):29–32

# Chapter 2
# Channel Forms and Patterns

**Abstract** *Channel forms and patterns* are important determinants of ease of movement of water and sediment and immediate clearance of materials from up slope. *Twenty five (25) cross-sections* have been drawn along the lower reach of the Rupnarayan River, by measuring river depth using *Echo Sounder* and collecting stuff readings using leveling instrument, to understand the channel forms and patterns. Most of the cross-sections, except two (at Kolaghat-AA$'$ and Geonkhali-YY$'$) are *asymmetrical* in nature. Depth is more near Kolaghat (maximum 10.5 m) and Geonkhali (maximum 11.5 m) than the middle of these two extremes (<8 m). Most of the portions near two banks and around the *mid channel bar* have the depth less than 2 m. *Width-depth ratio* is less near Kolaghat (38.09) and Geonkhali (130.43), but in the middle portion width-depth ratios are more than 300 and becomes 1111.10 in cross-section QQ$'$. Sudden expansion and widening of the channel near Kolaghat (width is 400 m at AA$'$, 1650 m at BB$'$, 2900 m at CC$'$) leads to *flow separation*, reduction of energy and deposition of sediment (during low tide) and near Geonkhali sudden constriction (bottle neck shape) of the channel (width-depth ratio—130.43) hinders free draining of ebb tide water leading to ponding effect and reduction of velocity and stream energy to drain the sediments and initiates sedimentation.

**Keywords** Channel forms and patterns · Asymmetry of channel · Width-Depth ratio · Flow separation · Sedimentation

## 2.1 Introduction

*Channel forms and patterns* are important determinants of ease of movement of water and sediment and immediate clearance of materials from up slope. Both the magnitude of discharge and character of the flow regime influence channel form (Harvey 1969; Stevens et al. 1975). Leopold and Maddock (1953) described the *morphology of river channels* in terms of the relationships between discharge (Q), width (W), depth (D) and velocity (V) and their rate of changes. Dury (1974), Wolman and Miller (1960) studied the relationship between discharge and channel

© The Author(s) 2018
S. Kumar Maity and R. Maiti, *Sedimentation in the Rupnarayan River*,
SpringerBriefs in Earth Sciences, https://doi.org/10.1007/978-3-319-62304-7_2

morphology. Schumm (1974) studied the relationship between channel patterns and sediment load and gradient in a flume. Generally width increases faster than depth and velocity in a downstream direction. Parker's (1978) model and Bagnold's (1980) equation imply that for a given discharge and slope, a larger sediment load requires a wide channel. Bagnold (1966) postulated a link between stream power and sediment transport capacity by suggesting that stream power provided an integrative parameter for relating sediment transport to channel hydraulics. Leopold and Maddock (1953) explained the concavity of the longitudinal profile in terms of the downstream decrease of sediment load in relation to discharge. Keller (1972) proposed a five stages model of the development of pools and riffles in an alluvial river. Both meander bends and pool-riffle sequences are repetitive, meander wave length and amplitude, as well as pool-riffle spacing are related to channel width and pools are associated with meander bends while cross-over with riffles. Langbein and Leopold (1966) hypothesized that the *meandering pattern* represents a least work tendency and an equalization of power expenditure. Lewin (1977) classified meanders pattern changes as autogenic or allogenic. Autogenic changes are those inherent in river regime, including changes made by channel migration, crevassing, neck cutoffs etc. Allogenic changes are those taking place in response to some outside influence such as climatic fluctuations or human activities etc.

## 2.2  Field Monitoring and Applied Methodology

A total number of *twenty five (25) cross-sections* (Fig. 2.1) have been drawn all along the lower reach of the Rupnarayan River to understand the nature of channel forms and patterns. In this regard, depth of river is measured by *Echo Sounder* interfaced with position fixing system (Fig. 2.2a). Stuff readings have been collected using leveling instrument considering several numbers of change points (Fig. 2.2b). The *bench mark* of 6.5 m near Kolaghat Thermal Power Project has been used as standard for the calculation of reduced levels at all the stations. Then, all the cross profiles have been drawn using these reduced levels. The long profile is drawn with the help of the *thalweg positions* of all the cross profiles. The spatial distributions of depth and *shoaled up area* along the river are also done by this methods along with the data collected by G.P.S. receiver. Cross-sectional area and width-depth relation have been estimated by the collected data in the field.

Slope of the river bed is calculated by the equation-

$$\theta = \tan^{-1}\frac{r}{h} \tag{2.1}$$

where, $\theta$ = the river bed slope, r = difference of reduced level at the first and last points and h = the horizontal distance between same two points.

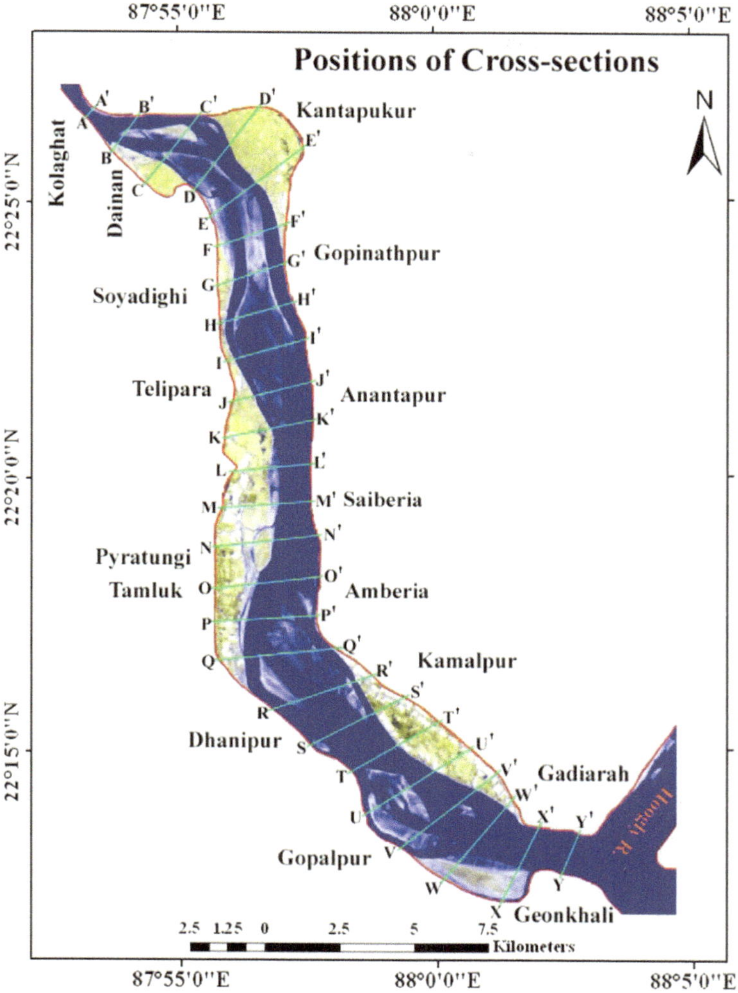

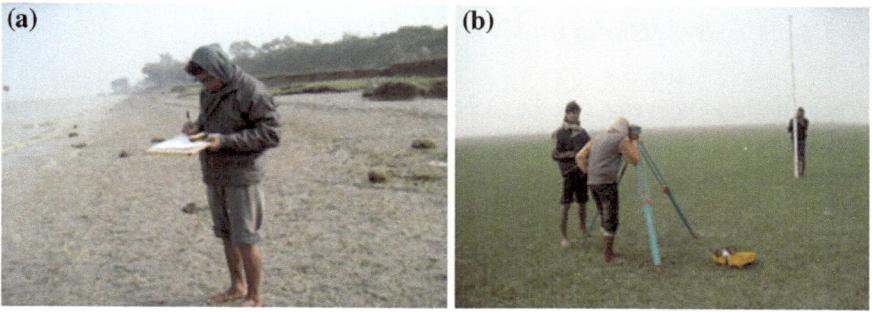

**Fig. 2.1** Location of cross-sections drawn in the study area

**Fig. 2.2** Collection of primary data by G.P.S. receiver (**a**) and leveling instrument (**b**)

*Sinuosity Index (SI)* is calculated following the equations-

$$SI = \frac{\text{Thalweg Length}}{\text{Valley Length}} \tag{2.2}$$

(Leopold and Wolman 1957).

$$SI = \frac{\text{Length of Channel}}{\text{Length of Meander belt axis}} \tag{2.3}$$

(Brice 1964).

## 2.3  Nature of Cross-Sections

Cross-sectional form of natural channels is characteristically irregular in outline, locally variable and is roughly parabolic in shape. The shape of cross-section of a river channel is a function of the flow, the quantity and character of the sediment and the composition of the materials making up the bed and banks of the channel (Knighton 1981). Width, mean depth, cross-sectional area, wetted perimeter, hydraulic radius, maximum depth, bed width, width-depth ratio etc. are the important parameters commonly used to describe channel form (Fahnestock 1963; Knighton 1981). In the study area only two cross-sections (one at Kolaghat-AA$'$ and another at Geonkhali-YY$'$) are symmetrical (the slope, length and configuration of both the sides of the cross-sections are same from the thalweg position) and remaining cross-sections are *asymmetric in nature* (Fig. 2.4). As the study area is facing the problem of rapid rate of sedimentation that's why the development and *shifting of shoal area*, re-distribution of sediments and *shifting of thalweg* position etc. are causing the cross-sections to be asymmetrical in nature. Spatial variation of erosion and deposition, river meandering are also contributing to make the channel asymmetrical (Middleton and Southard 1978). Asymmetry in the cross-profile is very common in areas of gently sloping beds (Knighton 1981).

## *2.3.1  Cross-Sections and Velocity Distribution*

The *distribution of velocity* varies according to the shape of the channel (Morisawa 1963). Where the channel is wide and shallow the velocity gradients are higher near the bed than at the banks. If the channel is narrow and deep, isovels (lines joining points of equal velocity) will be crowded near the banks and farther apart along the bed. Shear will then be greater at the banks than at the bed. These concepts are important in considering erosion of the channel and sediment transport through the reach (Hubbell and Matejka 1959). Where the channel is asymmetric, velocity distribution is skewed and, indeed, may be separated into two or more cells (Morisawa 1985). In the study area most of the cross-sections are asymmetrical in shape that leads to *flow separation*, *velocity separation* and reduction of velocity towards the shallower section. This results in sedimentation, whereas in deeper portion scouring is dominant (Maity and Maiti 2012). But, if the channel is symmetric the highest velocity is in the middle of the channel just below the surface (Knighton 1988).

## 2.3.2  *Cross-Sections and Turbulence*

The mechanics of river flow are greatly influenced by turbulence, i.e., the super-position of secondary movements on predominant downstream flow of water (Morisawa 1963). Leighley (1934) related *regions of turbulence* in a river cross-section to *realms of velocity*. He divided the transverse section into three parts-(1) An inner axial region of high velocity and moderate turbulence. (2) An outer realm of low velocity and low turbulence. (3) An area between the other two regions where turbulence is maximum and velocity is moderate (Fig. 2.3).

The size of the belts and zones of turbulence shift laterally and vertically with cross-section shape. In symmetrical form zone-1 is located at the middle of the cross-section and zone-2 and zone-3 are on both sides of it with same areal extension. But in asymmetric cross-section all the three zones are shifted towards one side (deeper portion) of it, with zone-2 and zone-3 on both sides of zone-1

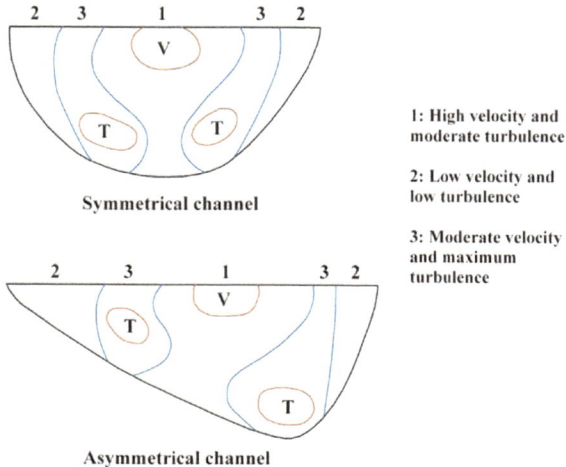

**Fig. 2.3** Relation between channel shape, velocity and turbulence (After Leighley 1934)

1: High velocity and moderate turbulence

2: Low velocity and low turbulence

3: Moderate velocity and maximum turbulence

Symmetrical channel

Asymmetrical channel

(Fig. 2.3). Here, the areal extension of zone-2 and zone-3 is more towards the shallower portion than the deeper portion of the cross-section (Leighley 1934).

As most of the cross-sections in the study area are asymmetrical, the *zone of low velocity and low turbulence (zone-2)* is wider towards the shallower section than the deeper section and creates favorable environments for sediment deposition. Sudden widening of the channel near Kolaghat leads to *flow separation*, reduction of energy and deposition of sediment (during low tide) and near Geonkhali the flow getting obstructed due to constriction of channel, causing reduction of flow velocity and stream energy and deposition of sediments (during low tide). In the middle portion, the sediments are redistributed during two phases of tide and deposited during low tide as low tide is longer and weaker (Maity and Maiti 2012).

## 2.4  Longitudinal Profile and Thalweg Profile of the Channel

*Long profile* of the stream bed is irregular because of variations in bed material size and shape, riffle-pool spacing, and other variables (Knighton 1984; Leopold 1992). The long profile of the study area represents that at two extreme ends (one at Kolaghat and another at Geonkhali) the width of the river is 400 m and 1500 m respectively but the maximum depth is 8.5 m and 11.5 m respectively (Fig. 2.4). In between these two extremes, the channel is wide enough with respect to the depth. Longitudinal profiles are important for identifying the general trends of down cutting and aggradations (Bernhardt et al. 2005). The slope of the longitudinal profile determines the energy of the water flow and provides a link between the long profile and the capacity for a reach to transport sediments (Brookes and Sear 1996).

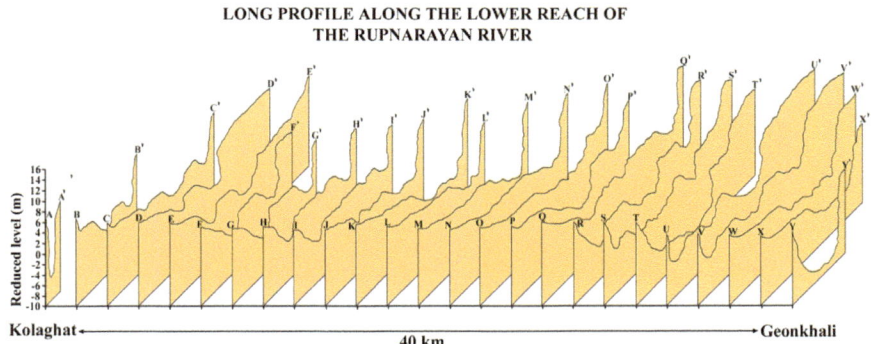

**Fig. 2.4** Long profile of the lower reach of the Rupnarayan River. (*Source* Field measurement)

Average gradient of the thalweg profile (between Kolaghat and Geonkhali) is 0.0000105, indicating very gentle slope of the region.

## 2.5 Depth Distribution along the Lower Reach

*Channel shape* depends on the distribution of depth across and along the channel. Distribution of depth across the channel depends largely on channel width, which depends on discharge. The mean depth of a stream varies greatly from reach to reach (along the river) depending on channel width, slope, volume of discharge and riffle-pool spacing etc. With an increase of volume of water discharge through a channel the depth will increase (Leopold and Maddock 1953). In the study area the longitudinal distribution of depth is highly dynamic in nature. Near Kolaghat and Geonkhali the depth is more compared to the middle section of the reach. Maximum depth (>10 m) is observed at the thalweg of the cross-sections (symmetrical valley) at Kolaghat and Geonkhali (Fig. 2.5). In the middle of these two extremes the depth is mainly <8 m. There are some scattered zones where scouring and shoaling on river bed lead to the formation of *irregularity in cross-section*. Most of the portions near two banks and around the *mid channel bar* have the depth less than 2 m (Fig. 2.5). Generally the thalweg line lies along the middle of the channel, except at Soyadighi and Gopinathpur, where the development of a mid-channel bar, the *diversion of the flow* is observed (Fig. 2.5).

### 2.5.1 Seasonal Variation of Depth Distribution

Distribution of depth along the lower reach is the result of the interaction between *upstream discharge* and *tidal discharge* in different seasons (Maity and Maiti 2012). The river discharge is impounded during flood period and released during ebb period. When the material carried by littoral drift reaches a river mouth during ebb period, it is picked up by the tidal current and is carried into the river channel by the flood current. The incoming high tide brings a lot of sediments from seaward side and deposits it on the bed of Rupnarayan River in absence of any upland discharge during the non-monsoon months (Rhodes 1950). As a result of continuous sedimentation the depth of river is decreased to a great extent. During Pre-monsoon and Post-monsoon, the maximum depth near Kolaghat region are 10.1 m and 10.5 m respectively and near Geonkhali the depth are 11.3 m and 11.5 m respectively. In the middle portion, all the places have maximum depth of <10 m. During freshet condition (July-September) the impact of high tide is felt to be reduced due to discharge of large volume of water from the upland area of the

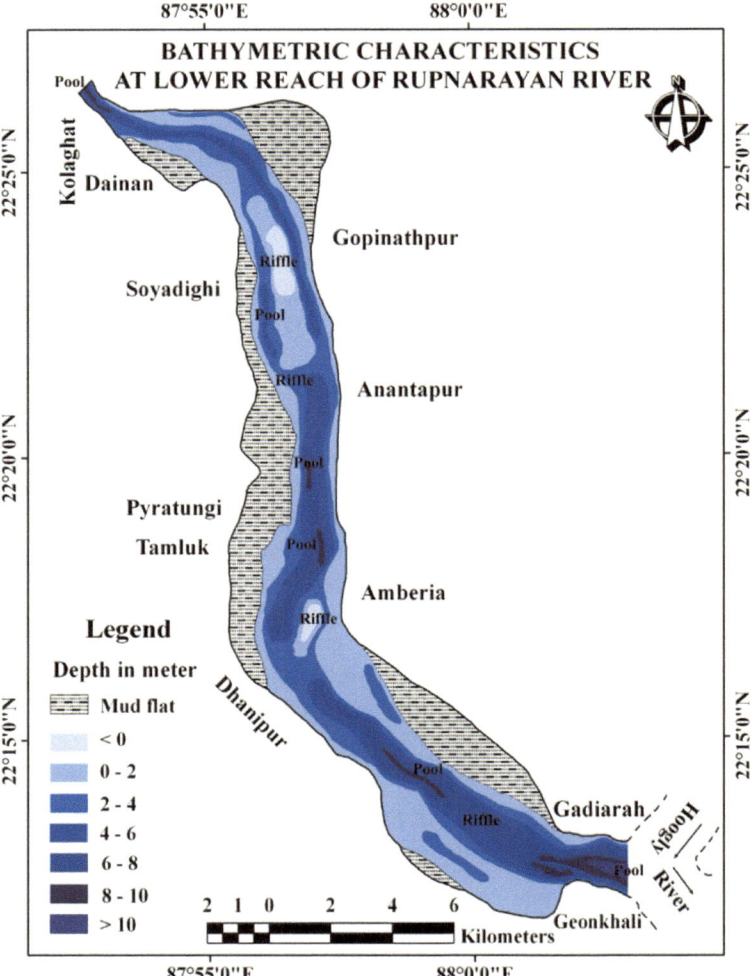

**Fig. 2.5** Distribution of depth along the lower reach of the Rupnarayan River. (*Source* Field measurement)

river. During monsoonal period scouring of river bed is more compared to the shoaling and so, the depth of river is increased (Maity and Maiti 2012). Maximum depth near Kolaghat and Geonkhali are 13.2 m and 14.9 m respectively in monsoon and in between these two extremes most of the places have maximum depth of more than 10 m (Table 2.1).

**Table 2.1** Seasonal variation of Depth distribution and Width-depth ratio

| Cross-section | Maximum Depth (m) | | | Average Width-depth ratio |
|---|---|---|---|---|
| | Pre-monsoon | Monsoon | Post-monsoon | |
| AA' | 10.1 | 13.2 | 10.5 | 38.09 |
| BB' | 5.1 | 7.6 | 4.5 | 366.66 |
| CC' | 4.8 | 7.8 | 5.4 | 659.09 |
| DD' | 5.1 | 8.2 | 5.4 | 666.66 |
| EE' | 5.0 | 7.3 | 4.2 | 916.66 |
| FF' | 6.1 | 7.5 | 4.6 | 543.47 |
| GG' | 5.5 | 8.7 | 5.6 | 419.64 |
| HH' | 7.1 | 8.8 | 5.9 | 440.47 |
| II' | 5.3 | 6.4 | 3.8 | 723.68 |
| JJ' | 6.2 | 8.0 | 5.2 | 528.84 |
| KK' | 6.1 | 8.3 | 5.9 | 525.42 |
| LL' | 8.8 | 11.7 | 8.5 | 323.52 |
| MM' | 7.3 | 9.3 | 6.8 | 455.88 |
| NN' | 8.1 | 10.5 | 8.6 | 389.53 |
| OO' | 5.7 | 9.3 | 6.1 | 590.16 |
| PP' | 6.6 | 10.7 | 7.5 | 446.66 |
| QQ' | 4.1 | 7.8 | 3.6 | 1111.10 |
| RR' | 7.1 | 9.8 | 7.3 | 495.89 |
| SS' | 7.4 | 9.6 | 7.5 | 480 |
| TT' | 6.1 | 10.4 | 6.1 | 410.97 |
| UU' | 4.7 | 10.8 | 6.5 | 500 |
| VV' | 6.8 | 10.4 | 6.7 | 572.22 |
| WW' | 8.2 | 11.5 | 7.8 | 461.53 |
| XX' | 8.2 | 11.8 | 9.4 | 305.31 |
| YY' | 11.3 | 14.9 | 11.5 | 130.43 |

(*Source* Field measurement)

## 2.6 Width-Depth Ratio

*Width-depth ratio* is one of the important parameter to understand the channel form. Relatively narrow and deep river with lower width-depth ratio is capable of moving more suspended sediment by turbulence and velocity of flowing water, but shallow wider channel, having large width-depth ratio, is more efficient to transport bed load by shear on the bottom of the stream (Coleman 1969). In the study area width-depth ratio is less near Kolaghat and Geonkhali (38.09 and 130.43 respectively), but in the middle portion width-depth ratios are more than 300. It becomes 1111.10 in the cross-section of QQ' (Table 2.1). The width-depth ratio changes continuously to adjust with the discharge, load type and slope of the bed. River discharge, rock type, slope, type of load being transported, type of sediment at the channel's perimeter, and rock uplift rate have all been shown to control river width and width-depth ratio (Schumm 1960). Channel perimeters with a high amount of coarse grained materials tend to be wide and shallow (Schumm 1960).

## 2.7  Channel Widening, Flow Separation and Diversification

There is a sudden *expansion and widening of the channel* near Kolaghat region which leads to flow separation and diversion of flow. The width of the cross-section AA$'$ is 400 m, but it suddenly increased to 1650 m at cross-section BB$'$ and again increased to 2900 m at CC$'$ at downstream of the Kolaghat (Figs. 2.1, 2.6 and 2.7a). In spite of this one, there are another two diversion and flow separation in between the region of Pyratungi and Dhanipur and Dhanipur and Geonkhali (Figs. 2.1, 2.6

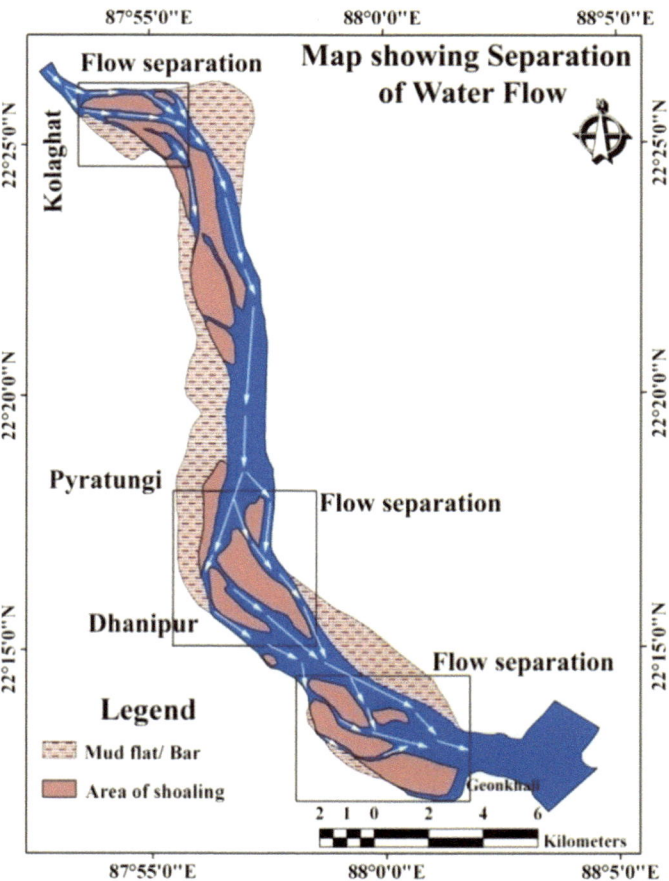

**Fig. 2.6** Separation of flow at lower reach of the Rupnarayan River

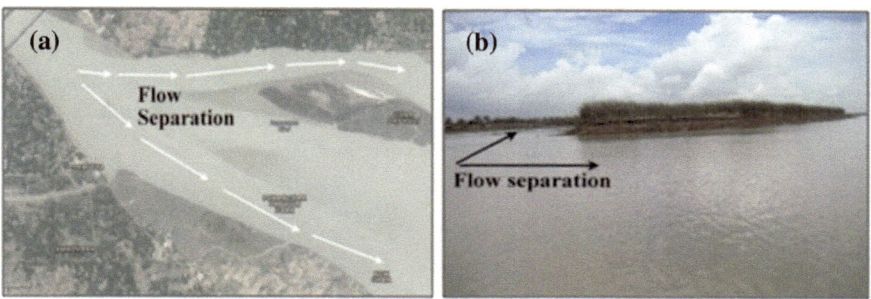

**Fig. 2.7**  Flow separation near Kolaghat (**a**) and Pyratungi (**b**)

and 2.7b). Flow separation and diversion of channel is caused where channel widens, gradient declines or discharge decreases (Naden and Brayshaw 1987). Smith (1974) mentioned that abrupt expansion of channel banks is important cause of flow separation.

## 2.7.1   *Flow Separation, Velocity Reduction and Sedimentation*

*Channel widening* and associated *flow separation* is important cause of reduction of flow velocity, reduction of stream energy and sedimentation on river bed. Bagnold (1966) noted that stream power is expressed as the ability of the water flow to do work within a channel in a particular instant of time and that rate of work indicates the availability of energy in a stream to transport the sediments. The rate of doing work by the available energy or the cross-sectional total stream power ($\Omega$), can be expressed as the combined effect of the specific weight of the fluid ($\gamma$), the discharge of water (Q) and the energy slope or the slope of the channel bed (S):

$$\Omega = \gamma QS$$

If there is any change in the magnitude of water flow, due to any reason (single diversion for example), the total available stream power will be changed. In case of river diversion and separation of flow the available stream power is reduced, causing the reduction of the ability of the flow of doing work at a certain rate, i.e., the *reduction of sediment transport power* (Letter Jr et al. 2008). Watson et al. (1999) pointed out that though the bed slope and sediment particle size remain constant but the water flow will decrease because of channel diversion and flow separation. It will reduce the sediment transport capacity of the river resulting into the sedimentation and aggradation on the river bed. Near Geonkhali region sudden constriction (bottle neck shape) of the channel (width-depth ratio—130.43) hinders the free draining of ebb tide water leading to ponding effect and velocity reduction that leads to sudden reduction of energy to drain the sediments. Again, during flood tide sudden widening at upstream of Geonkhali (width-depth ratio-130.43 at YY', 305.31 at XX', 461.53 at WW' and 572.22 at VV') leads to flow separation and energy reduction that also incapacitates the stream to favour *sedimentation*.

## 2.8  Stream Meandering

Channel pattern is usually represented by the ratio of stream length to valley length (stream sinuosity). As the asymmetry increases, flow impinges on the banks and erodes them to form sinuous pattern (Schumm 1963). The value of Sinuosity Index (SI = 1.35) between the reach Kolaghat and Gopinathpur represents the braided nature of the channel (the channel is divided by bars and small islands), whereas the channel is of Sinuous pattern (SI = 1.08) between Gopinathpur and Dhanipur reach. Between Dhanipur and Geonkhali, the channel is braided in nature (SI = 1.38). For the entire lower reach of the Rupnarayan River, the Sinuosity Index (SI = 1.31) indicates the *braided pattern* of the channel. The gradient of the river bed is very gentle to almost flat (0.0000105) and the mean grain size is fine sand to coarse silt in nature, which have an impact to control channel pattern. Generally speaking, as gradient and particle size decreases, there is a corresponding increase in sinuosity. Braided pattern of the channel leads to flow separation, reduction of stream energy, shifting of the area of scouring and shoaling and changing of the river course.

## References

Bagnold RA (1966) An approach to the sediment transport problem from general physics. US Geol Survey Prof Paper, 422:37

Bagnold RA (1980) An empirical co-relation of bed load transport rates in flumes and natural rivers. Proc R Soc 372A:453–473

Bernhardt ES, Palmer MA, Allan JD et al (2005) Synthesizing U.S. river restoration efforts. Science 308:636–637

Brice JC (1964) Channel patterns and terraces of the Loup River in Nebraska. US Geol Survey Prof Paper, 422D

Brookes A, Sear DA (1996) Geomorphological principles for restoring channels. In: Brookes A, Shields J (eds) River channel restoration: guiding principles for sustainable projects. Wiley, Chichester, pp 75–101

Coleman JM (1969) Brahmaputra River: channel process and sedimentation. Sediment Geol 3:129–139

Dury GH (1974) Magnitude—frequency analysis and channel morphometry. In: Morisawa ME (ed) Fluvial geomorphology. SUNY-Binghumton, New York, pp 9–21

Fahnestock RK (1963) Morphology and hydrology of a glacial stream-White River, Mount Rainier, Washington. US Geol Surv Prof Pap, 422A

Harvey AM (1969) Channel capacity and the adjustment of streams to hydrologic regime. J Hydrol 8:82–98

Hubbell DW, Matejka DQ (1959) Investigations of sediment transportation Middle Loup River at Dunning, Nebraska. US Geol Surv Water Suppl Pap 1476:123

Keller EA (1972) Development of alluvial stream channels: a five stage model. Bull Geol Soc Amer 83:1531–1536

Knighton AD (1981) Local variations of cross-sectional form in a small gravel bed stream. J Hydrol (New Zealand) 20:131–146

Knighton AD (1984) Indices of flow asymmetry in natural streams: definition and performance. J Hydrol 73:1–19

Knighton AD (1988) The impact of the Parangana Dam on the River Mersey, Tasmania. Geomorphology 1:221–237

Langbein WB, Leopold LB (1966) River meanders—theory of minimum variance. US Geol Surv Prof Pap 422:1–15

Leighly JB (1934) Turbulence and the transportation of rock debris by streams. Geogr Rev 24:453–464

Leopold L (1992) Sediment size that determines channel morphology. Dynamics of Gravel Bed Rivers. Wiley, New York, p 15

Leopold LB, Maddock T (1953) The hydraulic geometry of stream channels and some physiographic implications. US Geol Surv Prof pap, p 252

Leopold LB, Wolman MG (1957) River channel patterns-braided, meandering and straight. US Geol Surv Prof pap 282B:39–85

Letter Jr JV, Pinkard Jr CF, Raphelt NK (2008) River diversion and shoaling. Coastal and Hydraulic Engineering Technical Note, ERDC/CHL CHETN-VII-9, Vicksburg, MS: US Army Engineer Research and Development Center

Lewin J (1977) Channel pattern changes. In: Gregory KJ (ed) River channel changes. Wiley, New York, pp 167–184

Maity SK, Maiti RK (2012) Impact of sedimentation on development and shifting of shoal area, pools and riffles and thalweg position at lower reach of Rupnarayan River-A case study. Indian J Power River Val Dev 24:46–54

Middleton GV, Southard JB (1978) Mechanics of sediment transport. SEPM Short Course No 3, Binghamton, New York, p 248

Morisawa ME (1963) Distribution of stream-flow direction in drainage patterns. J Geol 71(4):528–529

Morisawa M (1985) Rivers: forms and process. Longman Inc, New York

Naden PS, Brayshaw AC (1987) Small and medium scale bed forms in gravel-bed Rivers. In: Richards KS (ed) River channels: environment and process. Blackwell, Oxford, pp 249–271

Parker G (1978) Self-formed straight rivers with equilibrium banks and mobile bed. Part 1: the sand-silt river. J Fluid Mech 89(1):109–125

Rhodes RF (1950) Effects of salinity on current velocities, US Corps of Engineers, Committees Tidal Hydraulics, report No 1, p 94

Schumm SA (1960) The shape of alluvial channels in relation to sediment type. US Geol Surv Prof Pap, 352-D

Schumm SA (1963) A tentative classification of alluvial river channels. US Geol Survey Circular, p 477

Schumm SA (1974) Geomorphic thresholds and complex response of drainage systems. In: Morisawa ME (ed) Geomorphology. SUNY, Binghamton, pp 299–310

Smith KG (1974) Erosional processes and landforms in Dadland national monument, South Dakota. Geol Soc Amer Bull 69:975–1008

Stevens MA et al (1975) Non-equilibrium river form. J Hydr Div, ASCE 101:557–566

Watson CC, Biedenharn DS, Scott SH (1999) Channel rehabilitation: processes, design and implementation. Workshop proceedings, US Army Engineer Research and Development Center, Coastal and Hydraulics Laboratory, July 1999

Wolman MG, Miller J (1960) Magnitude and frequency of forces in geomorphic processes. J Geol 68:54–74

# Chapter 3
# Stream Hydraulics

**Abstract** Seasonal variation of stream hydraulics was monitored to understand and explain the *mechanism of sedimentation*. Water velocity is measured by floating method and using digital current meter. Density of water is measured by testing the collected water samples in the laboratory. The *pattern and nature of flow* is identified by *Reynolds Number* and *Froude Number*. Spatial and temporal variation of stream depth and velocity indicates the unsteady and non-uniform nature of flow. Values of Reynolds Number are >6248.66 in all the cross-sections, which indicates the *turbulent pattern of flow*. Values of Froude Number (0.2493–0.5238), indicates the *sub-critical nature of flow*. Maximum surface water velocity is measured as 1.22–2.02 m/s in pre-monsoon, 1.42–2.25 m/s in monsoon and 1.33–2.05 m/s in post-monsoon season. High tide water velocity is more (0.87–2.25 m/s) than the velocity in low tide (0.65–2.05 m/s). Stronger high tide transport more sediment towards upstream than that discharged towards downstream during low tide, causing *sedimentation*. During dry season paucity of rainfall causes *less discharge of water* (850–4160 m$^3$/s), reduction of stream energy and sediment transporting capacity which allows sedimentation. But in monsoon season occurrence of huge rainfall increases the water discharge (3455–9050 m$^3$/s), stream energy, sediment transporting capacity and reduces sedimentation rate.

**Keywords** Sedimentation · Water velocity · Flow pattern · Water discharge

## 3.1 Introduction

Proper understanding of the distribution of different point hydraulic variables (water depth, velocity or shear stress etc.) in natural streams is very important to the hydraulic engineers, fluvial geomorphologists and stream ecologists (Chiu and Tung 2002; Mérigoux et al. 2009; Rosenfeld et al. 2011; Grandgirard et al. 2013). Deterministic numeric models and Statistical hydraulic models are commonly used

© The Author(s) 2018
S. Kumar Maity and R. Maiti, *Sedimentation in the Rupnarayan River*,
SpringerBriefs in Earth Sciences, https://doi.org/10.1007/978-3-319-62304-7_3

for the prediction and mapping of hydraulic patterns within reaches (Legleiter et al. 2011). Relations of *hydraulic geometry* suggest that rates of increase of width and velocity are greater towards downstream and lesser for depth than average values for every river in the world (Leopold et al. 1964; Wolman and Gerson 1978; Knighton 1998). Leopold and Maddock (1953) considered the channel properties as continuous functions of increasing discharge towards downstream and introduced hydraulic geometry approach. Velocity of water flow decreases over bars due to higher roughness and it facilitates the deposition of sediments on the leeward sides of streams (Nepf 1999). Due to the difference of vertical velocity in the flow the bottom layer of water moves slowly in a slightly sinuous pattern, but the upper layer water flow faster and straighter (Gorycki 1973). At bends, velocity distribution is asymmetrical, shear stress is high, resistance to flow is high and surface roughness is low (Hickin 1978). Nordin (1963) mentioned that maximum velocity generally occurring near, but somewhat below the water-surface and it varies logarithmically in vertical profile. Graphical representation of vertical distribution of velocity shows that velocity increases from the bed to the water surface (Savini and Bodhaine 1971). According to Richards (1973) and Knighton (1979) as depth increases with discharge at a cross-section, the effect of grain roughness is drowned out and flow resistance decreases and velocity may also tend to change more slowly, producing non-linearities in hydraulic geometry.

## 3.2  Field Monitoring and Applied Methodology

*Water velocity* (surface water velocity and water velocity at different depths) is measured by floating method and using digital current meter (Fig. 3.1). Density of water is measured by collecting water samples and testing it in the laboratory of Indian Institute of Technology (IIT), Kharagpur. Tidal water level data have been collected at different gauge stations at an interval of one hour, for the calculation of *water discharge*. Discharge of water has been calculated for all the months in a year

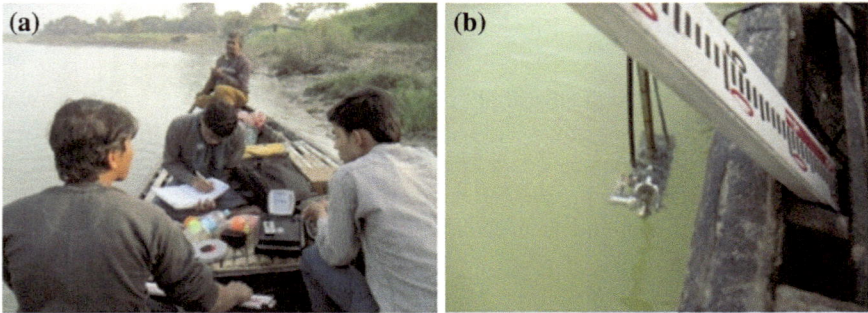

**Fig. 3.1**  Measurement of water depth by echo-sounder (**a**) and water velocity by current meter (**b**)

and then the average discharge is calculated for pre-monsoon (February to May), monsoon (June to September) and post-monsoon (October to January) seasons.

The pattern of flow is identified by *Reynolds Number* (Reynolds 1874)-

$$Re = \rho \frac{VR}{\mu} \tag{3.1}$$

Where, Re = Reynolds Number, V = mean velocity of flow, R = hydraulic radius, $\mu$ = dynamic viscosity, measured in the laboratory. Re < 500 indicates laminar flow, Re = 500–2000 indicates transitional flow and Re > 2000 indicates the flow is turbulent.

The nature of flow is identified by *Froude Number*-

$$Fr = \frac{v}{\sqrt{gD}} \tag{3.2}$$

Where, Fr = Froude Number, v = mean velocity of water, g = gravitational constant, $D$ = water depth.

Fr < 1 indicates sub-critical flow, Fr = 1 indicates critical flow and Fr > 1 indicates the flow is super critical in nature.

Water discharge is calculated following Leopold and Maddock (1953)-

$$Q = WDV \tag{3.3}$$

Where, Q = the discharge, W = the width, D = depth and V = mean velocity.

## 3.3 Steady and Unsteady or Uniform and Non-uniform Flow of Water

In *steady flow* the depth and velocity do not vary in magnitude or direction with respect to time. The flow becomes *unsteady* if the depth and velocity fluctuates in terms of magnitude or direction with time. When there is no change of depth and velocity with distance along the stream, the *flow is uniform* (the gravitational and frictional components are in balance). If depth and velocity change with distance along a channel, *flow is non-uniform* or varied (Morisawa 1985). The study area is characterized by great temporal and spatial variation of depth and velocity, which indicates *the unsteady and non-uniform flow* of water (Maity and Maiti 2012). At Kolaghat (cross-section-AA′) the maximum depths are 10.1 m and 10.5 m during pre-monsoon and post-monsoon respectively, while it is 11.3 m and 11.5 m respectively near Geonkhali (cross-section-YY′). But in the middle portion the maximum depth is <10 m at all the places (Table 2.1). During monsoon season all the places exhibit greater depth than pre-monsoon and post-monsoon seasons. Maximum depth near Kolaghat and Geonkhali region are 13.2 m and 14.9 m respectively in this season

**Table 3.1** Spatial and temporal variation of velocity distribution

| Cross-sections | Maximum Surface Water Velocity (m/sec) | | |
|---|---|---|---|
| | Pre-monsoon | Monsoon | Post-monsoon |
| AA' | 1.74 | 2.10 | 1.81 |
| BB' | 1.69 | 1.78 | 1.65 |
| CC' | 1.48 | 1.56 | 1.50 |
| DD' | 1.50 | 1.68 | 1.55 |
| EE' | 1.47 | 1.55 | 1.45 |
| FF' | 1.43 | 1.44 | 1.60 |
| GG' | 1.22 | 1.42 | 1.42 |
| HH' | 1.44 | 1.61 | 1.35 |
| II' | 1.48 | 1.59 | 1.65 |
| JJ' | 1.90 | 2.04 | 1.78 |
| KK' | 1.68 | 1.82 | 1.77 |
| LL' | 1.70 | 1.80 | 1.58 |
| MM' | 1.47 | 1.66 | 1.55 |
| NN' | 1.50 | 1.72 | 1.54 |
| OO' | 1.45 | 1.55 | 1.42 |
| PP' | 1.40 | 1.62 | 1.52 |
| QQ' | 1.48 | 1.59 | 1.41 |
| RR' | 1.70 | 1.83 | 1.85 |
| SS' | 1.73 | 1.90 | 1.77 |
| TT' | 1.48 | 1.64 | 1.41 |
| UU' | 1.30 | 1.48 | 1.33 |
| VV' | 1.84 | 2.01 | 1.82 |
| WW' | 1.88 | 2.13 | 1.90 |
| XX' | 1.96 | 2.15 | 2.01 |
| YY' | 2.02 | 2.25 | 2.05 |

(*Source* Field measurement)

(Table 2.1). The surface velocity of water is also variable at different places from upstream to downstream and in different seasons. Maximum surface water velocity in pre-monsoon (2.02 m/s) and post-monsoon (2.05 m/s) seasons are measured at Geonkhali (cross-section- YY') while minimum velocity (1.22 m/s and 1.33 m/s) are measured at cross-section GG' and UU' respectively. But during monsoon season the maximum and minimum velocity are 2.25 m/s and 1.42 m/s respectively (Table 3.1). Usually natural stream flow varies to compensate and adjust at bends, contractions and expansions, obstructions etc. Stream flow can change quickly with time or space, such as in a hydraulic jump or hydraulic drop (Morisawa 1985).

## 3.4   Pattern of Flow

*Reynolds Number* has been calculated at all the cross-sections along the lower reach of the Rupnarayan *River* (Fig. 2.1) to identify the *pattern of flow*. The values of Reynolds Number at all the stations range from 6248–15,927, indicating the turbulent flow pattern in the study area (Table 3.2). In turbulent condition the flow is chaotic, being characterized by mixing of water masses with exchange of turbulent energy from one water mass to another. Eddies and other secondary flow patterns are superimposed on the primary downstream flow (Morisawa 1985). At cross-sections AA′, FF′, KK′, LL′, NN′, OO′, PP′, QQ′, XX′ and YY′, the flow is more turbulent (values of Reynolds Number are 12268.53, 10254.51, 10617.92, 10542.51, 10728.53, 11381.71, 11286.52, 10730.39, 12439.64 and 15927.91 respectively), compared to other cross-sections, where the values are <10,000 (Table 3.2). The variation of sedimentation rate at different cross-sections is due to the variation of degree of turbulence, because the turbulence helps the river to lift heavier objects off the river bed and transport them downstream (Knighton 1998). In very few channels water flow smoothly, generally obstacles on the river bed cause swirling vortices to form. In a dominant downstream flow, chaotic swirls develop. Turbulent flow occurs in complex winding channels and in rivers with riffle and pool sequences (Knighton 1998). The long profile of the study area is characterized by sharp variation in depth, leading to *turbulence of flow* (Maity and Maiti 2012).

**Table 3.2** Reynolds Number and Froude Number

| Cross-sections | $Re = \rho \dfrac{VR}{\mu}$ | $Fr = \dfrac{V}{\sqrt{gD}}$ |
|---|---|---|
| AA′ | 12268.53 | 0.3429 |
| BB′ | 7328.92 | 0.3015 |
| CC′ | 6248.66 | 0.2767 |
| DD′ | 8265.28 | 0.2971 |
| EE′ | 9163.90 | 0.3178 |
| FF′ | 10254.51 | 0.3208 |
| GG′ | 8361.62 | 0.3009 |
| HH′ | 9288.38 | 0.2493 |
| II′ | 7418.55 | 0.2751 |
| JJ′ | 8910.17 | 0.2809 |
| KK′ | 10617.92 | 0.3318 |
| LL′ | 10542.51 | 0.3187 |
| MM′ | 9820.63 | 0.3017 |
| NN′ | 10728.53 | 0.3185 |
| OO′ | 11381.71 | 0.3419 |
| PP′ | 11286.52 | 0.3170 |
| QQ′ | 10730.39 | 0.3295 |

(continued)

**Table 3.2** (continued)

| Cross-sections | $Re = \rho\dfrac{VR}{\mu}$ | $Fr = \dfrac{v}{\sqrt{gD}}$ |
|---|---|---|
| RR′ | 8529.71 | 0.2830 |
| SS′ | 8219.64 | 0.2731 |
| TT′ | 8196.33 | 0.2948 |
| UU′ | 9730.83 | 0.3178 |
| VV′ | 9319.52 | 0.3319 |
| WW′ | 9318.60 | 0.3181 |
| XX′ | 12439.64 | 0.3987 |
| YY′ | 15927.91 | 0.5238 |

(*Source* Field measurement)

## 3.5  Nature of Flow

The influence of gravity on water flow can be identified by the value of *Froude Number*. The *nature of flow* at lower reach of the Rupnarayan River is identified by calculating the value of Froude Number at all the cross-sections (Fig. 2.1). At all the stations the values of Froude Number are <1, which indicates the *sub-critical flow* of water (Table 3.2) (Morisawa 1985). Highest and lowest value is estimated at the cross-section YY′ (0.5238) and HH′ (0.2493) respectively (Table 3.2). So, flow tends to be critical at YY′ and more sub-critical at HH′. Larger value of Froude Number (more critical flow) at YY′ leads to scouring of river bed compared to shoaling (Maity and Maiti 2012). Higher value of Froude number at YY′ is due to channel constriction. Knighton (1998) mentioned that in most of the cases the flow in natural stream is generally *sub-critical in nature*. So, the lower reach of the Rupnarayan River, the region under study is characterized by *subcritical* and *turbulent pattern of flow*.

## 3.6  Velocity Distribution

The distribution and *variation of velocity* in channel is important because it influences the process of erosion, transportation and deposition (Allan 1995). Being affected by tidal phenomenon, the lower reach of the Rupnarayan river exhibits considerable spatial and temporal variation of water velocity during high and low tide. High tide velocity is more compared to that in low tide. The distribution of velocity along and across the channel is affected by distribution of depth, river bed gradient, roughness of bed, meandering pattern, water discharge, location of mid-channel bar and pool-riffle sequence etc. Generally, maximum surface water velocity is measured along the *thalweg position*, while the portions of the channel near the banks have minimum velocity due to more friction and resistance force

(Dingman 1989). The *separation of flow* also causes the reduction of water velocity. Knighton (1998) mentioned that velocity is one of the most sensitive and variable property of open-channel flow because it depends on many other factors like, distribution of depth, shape of the channel, slope of the river bed, solute diffusion, flow resistance, roughness of the bed and sediment transport etc. Boundary condition also affects the velocity distribution causing turbulence and greater velocity variation at the stream bed and banks. In rough boundaries the velocity profile is a function of the size of the grains on the bed and depth of flow (French 1986).

### 3.6.1   Surface Water Velocity Distribution across the Channel

Mean velocity at a cross-section is very pertinent measure for defining the initiation of motion of sediments (Knighton 1998). Velocity of water increases towards the middle of a stream due to the reduction of frictional effect of the channel banks. Symmetrical distribution of water velocity across the channel will be disturbed due to the changes of shape and alignment of the channel, ultimately becoming typically asymmetric at channel bends. In asymmetrical channel, *velocity distribution is highly skewed* and may even be separated into two or more cells (Morisawa 1985). At Kolaghat and Geonkhali (AA' and YY'), cross-sections are symmetrical and so, have *non-skewed distribution of velocity*. Maximum surface water velocity is measured at the center of the cross-sections and it gradually decreases towards both the banks (during both high and low tide). But in case of all asymmetrical cross-sections and at bends the velocity distribution is asymmetric in nature. Here, around the shoaled up area and mid-channel bar the velocity is less and maximum velocity is measured towards one side of the cross-section, where depth is maximum (Figs. 3.2 and 3.3).

### 3.6.2   Surface Water Velocity Distribution along the Channel

Though there is a declining tendency of bed slope along most of the rivers but velocity tends to remain unchanged or increase slightly as the channel becomes hydraulically more competent and the resistance decreases towards the downstream direction. Spatial variation of velocity distribution is found along the studied reach depending on the sudden widening and contraction of channel, variation of depth, presence of mid-channel bar, bending of river, frictional and resistance force and variation of discharge etc. (Figs. 3.2 and 3.3). Bathurst (1993) mentioned that variations in the rate of change of velocity do occur along the river, since velocity is merely one variable that can be adjusted to accommodate the downstream increase in discharge. As the study area is characterized by very gentle slope (average

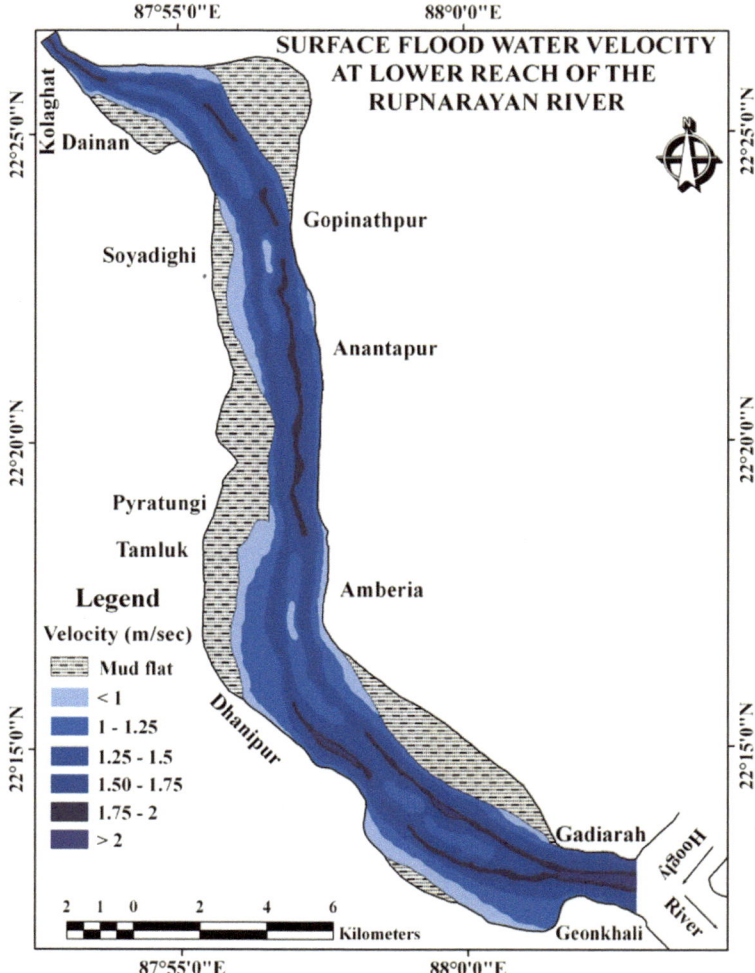

**Fig. 3.2** Distribution of surface velocity during high tide

gradient of the long profile is 0.0000105), the velocity distribution along the river is not largely affected by bed slope rather it is influenced by other factors.

### 3.6.2.1   Surface Water Velocity Distribution during High Tide

The lower reach of the Rupnarayan River is dominated by *semi-diurnal tide*, which is asymmetrical in nature. The *high tide is shorter and stronger* and so, velocity during high tide is more. Zones of surface water velocity distribution is not continuous along the river, rather these are discrete cells due to sudden channel

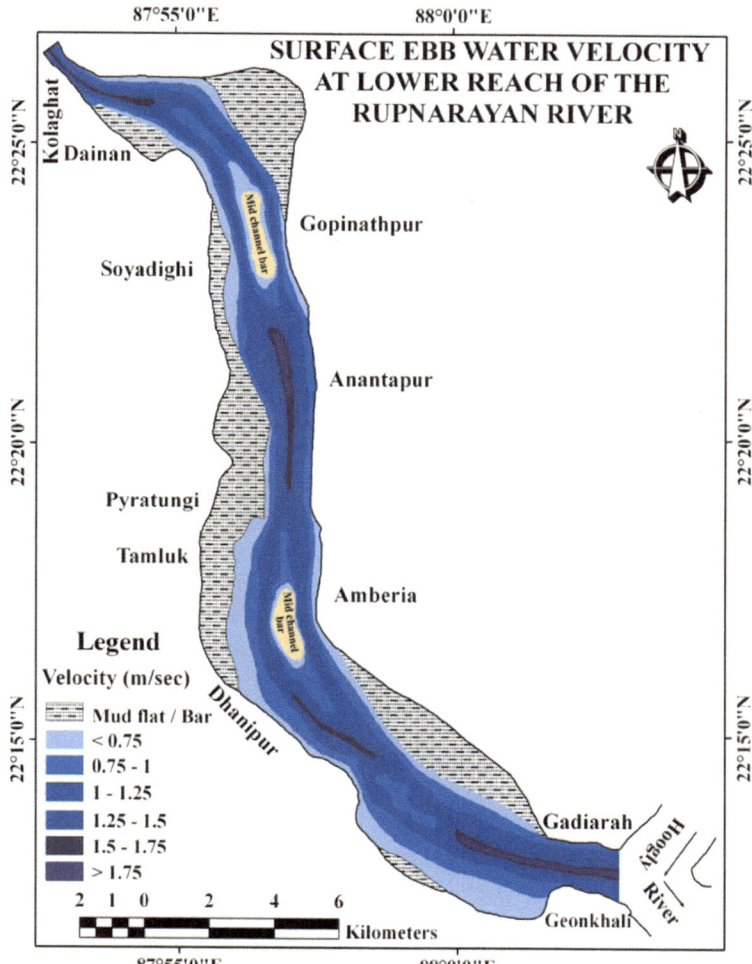

**Fig. 3.3** Distribution of surface velocity during low tide

widening and flow separation, obstruction of flow by mid-channel bar, bending of
the river etc. (Fig. 3.2). Maximum Surface water velocity near Geonkhali is measured as >2 m/s, but after that due to sudden widening of the channel (with respect
to high tide) the flow has been separated into two different cells declining the
maximum velocity to 1.75–2 m/s (Fig. 3.2). Between Pyratungi and Anantapur
reach the maximum surface water velocity (1.75–2 m/s) is measured along the
middle of the channel (symmetric distribution of velocity), but after that due to
development of mid channel bar the maximum velocity has reduced to <1.75 m/s.
Near Kolaghat region the maximum velocity (1.75 m/s and 2 m/s) is measured

along the middle of the channel. Most of the regions near two banks and that around and over the mid-channel bars are characterized by velocity of <1 m/s (Fig. 3.2).

### 3.6.2.2 Surface Water Velocity Distribution during Low Tide

The *surface water velocity* during low tide is less compared to high tide due to longer duration of the former. Velocity zones are discontinuous due to sudden channel widening and flow separation, obstruction of flow by mid-channel bar, bending of the river etc. (Fig. 3.3). Near Kolaghat region maximum velocity (1.5–1.75 m/s) is measured along the middle of the channel and gradual reduction of velocity towards the banks indicates *symmetrical velocity distribution*. Due to bending of river and development of mid-channel bar near Gopinathpur and Soyadighi reach, the flow has been separated into two cells causing the reduction of velocity (Fig. 3.3). Around the mid-channel bar velocity of water is minimum (<0.75 m/s) and maximum velocity is 1.25–1.50 m/s in the deeper section of the channel. Between Anantapur and Pyratungi reach maximum velocity (>1.5 m/s) is measured along the middle of the channel. Between Pyratungi and Dhanipur reach sudden widening of channel and the development of mid-channel bar leads to flow separation and reduction of velocity (Maximum velocity is <1.25 m/s) (Fig. 3.3). Near Geonkhali reach maximum velocity is >1.75 m/s along the middle of the channel and the velocity is reduced to 0.75–1 m/s near two banks.

## 3.6.3  *Relation between Depth and Velocity*

Velocity, velocity gradient and velocity distribution in a channel are affected by depth. Generally, the mean velocity in a cross-section varies inversely as the depth from water surface. The distribution of depth and velocity provide a simple, efficient and comprehensive approach to describing hydraulic variations in stream channels (Lamouroux et al. 1995; Lamouroux 1998). The time-averaged velocity usually increases from zero at the bed to the free-stream velocity at the edge of the boundary layer. An inner layer extending over 10–20% of the depth, in which velocity varies semi logarithmically with depth; and an outer layer in the upper 80–90%, where large scale turbulence is dominant and the velocity profile diverges from a semi logarithmic form. In the study area velocity is measured by Digital Current meter at different depths in three seasons. During monsoon, maximum velocity (2.48 m/s) is measured at 0.05 m depth while velocity is 0.09 m/s at the depth of 2.1 m (Fig. 3.4). During pre-monsoon and post-monsoon maximum velocity (2.05 m/s and 2.18 m/s respectively) is measured at the depth just below the water surface. Minimum velocity (0 m/s and 0.08 m/s) is measured at the depth of 2.5 m and 2.1 m in pre-monsoon and post-monsoon respectively (Fig. 3.4).

In all the seasons the relation between depth and water velocity is *moderately negative* (r = −0.556 in pre-monsoon, r = −0.595 in monsoon and r = −0.589 in

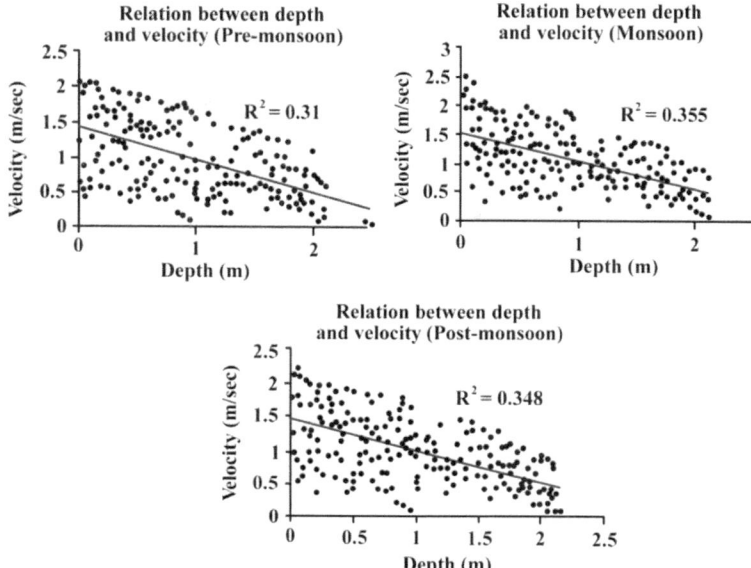

**Fig. 3.4** Relation between depth and velocity in different seasons

post-monsoon) (Fig. 3.4). The rate of reduction of velocity with depth is variable at different places in the reach in different seasons. The concepts of channel shape and velocity distribution are important in considering erosion, deposition and sediment transport through the reach (Hubbell and Matejka 1959).

### 3.6.4    Seasonal Fluctuation of Velocity

As the study area is affected by the typical monsoonal type of climate, there is a great *variation of riverine discharge* and velocity in pre-monsoon, monsoon and post-monsoon season. During lean period (pre-monsoon and post-monsoon) the velocity of water is reduced. Maximum surface velocity is measured as 2.02 m/s and 2.05 m/s in pre-monsoon and post-monsoon respectively (Table 3.1). During monsoon season *voluminous riverine discharge* increases the water velocity to 2.25 m/s. At all the cross-sections the velocity is more in monsoon than in pre-monsoon and post-monsoon season (Table 3.1). The increase of discharge tends to drown out roughness elements in the bed and thereby produce an increase in velocity. But the effect is not uniform and the velocity exponent 'm' (rates of change of velocity) can vary considerably from section to section (Park 1977).

### 3.6.5  Relation between Stream Velocity and Sedimentation

There is a close relationship between *velocity distribution* and *cross-sectional shape* to the erosional and depositional tendency in a channel (Stewardson 1999). In the study area the difference of velocity during high and low tide is the main reason behind the rapid rate of *sedimentation* (Maity and Maiti 2012). The velocity of water is more during high tide than in low tide. So, the available energy and sediment transporting capacity is more in high tide than in low tide condition. If, as in the situation above, flood periods are stronger, more sediment can be carried inland by flood tide than that can be drained by weaker ebb (assuming adequate sediment supply). So, more sediment penetrates inward during high tide but discharge of sediment is less during low tide and the region experiences a *net input of sediment*. The seasonal variation of rate of sedimentation (high in pre-monsoon and post-monsoon than in monsoon season) (Table 1.2) is due to the fluctuation of water discharge in different seasons. During pre-monsoon and post-monsoon season the riverine discharge is less, causing the reduction of velocity and hence the stream energy, giving the opportunity of the settlement of the sediments. But in monsoon season voluminous riverine discharge increases water velocity and stream energy, resulting the scouring of river bed instead of shoaling and thus the sedimentation rate is low.

## 3.7  Water Discharge

*Discharge of water* is the dominant independent variable controlling the depth, width, velocity of water and shape of the channel (Leopold and Maddock 1953). Adjustment of a channel, both at-a-station and downstream is mainly governed by the seasonal fluctuation of water discharge (Richards 1973). The discharge of water at lower reach of the Rupnarayan River is affected by the seasonal variation of rainfall in the upper catchment area and by the water, penetrating during high tide. Month wise variation of water discharge in different places is shown in Table 3.3. In the months of monsoon season, due to huge rainfall, water discharge during low tide is more compared to pre-monsoon and post-monsoon seasons. Mean maximum (9050 m$^3$/s) and minimum (3455 m$^3$/s) discharge in monsoon low tide is recorded at Geonkhali and Kolaghat respectively (Table 3.3). During dry season (pre-monsoon and post-monsoon) mean maximum and minimum low tide discharge are 4160 m$^3$/s (at Geonkhali) and 850 m$^3$/s (at Kolaghat) respectively (Table 3.3). High tide discharge in all the seasons is more than low tide discharge. In monsoon season, mean maximum (10,210 m$^3$/s) and minimum (4125 m$^3$/s) high tide discharge is measured at Geonkhali and Kolaghat respectively. But in dry season mean maximum and minimum high tide discharge are 7500 m$^3$/s and 1550 m$^3$/s at Geonkhali and Kolaghat respectively (Table 3.3). Discharge of water controls the stream energy, sediment transporting capacity and rate of

**Table 3.3** Spatial and seasonal variation of flood and ebb water discharge

| Month | Water discharge (m³/s) | | | | | | | | | | | |
|---|---|---|---|---|---|---|---|---|---|---|---|---|
| | Kolaghat | | Soyadighi | | Anantapur | | Pyratungi | | Dhanipur | | Geonkhali | |
| | High tide | Low tide | High tide | Low tide | High tide | Low tide | High tide | Low tide | High tide | Low tide | High tide | Low tide |
| February | 1720 | 760 | 2210 | 1120 | 2950 | 1400 | 2760 | 1690 | 4460 | 2750 | 7780 | 3970 |
| March | 1400 | 880 | 1860 | 1300 | 2500 | 1520 | 2690 | 1680 | 4230 | 2650 | 7720 | 3640 |
| April | 1430 | 820 | 2010 | 1270 | 2510 | 1400 | 2610 | 1740 | 4200 | 2700 | 7260 | 3550 |
| May | 1650 | 940 | 2120 | 1310 | 3040 | 1320 | 2940 | 1890 | 4310 | 2900 | 7240 | 3840 |
| **Average (Pre-monsoon)** | **1550** | **850** | **2050** | **1250** | **2750** | **1410** | **2750** | **1750** | **4300** | **2750** | **7500** | **3750** |
| June | 4000 | 3300 | 4810 | 4180 | 5070 | 4060 | 5300 | 4200 | 7540 | 5600 | 10,100 | 9210 |
| July | 4140 | 3540 | 4870 | 4000 | 5450 | 4170 | 5230 | 4390 | 7120 | 5740 | 10,480 | 9000 |
| August | 4350 | 3580 | 4940 | 4020 | 4760 | 4100 | 5010 | 4300 | 6800 | 5680 | 10,300 | 9010 |
| September | 4010 | 3400 | 4860 | 3900 | 4680 | 4150 | 5060 | 3950 | 6340 | 5580 | 9960 | 8980 |
| **Average (Monsoon)** | **4125** | **3455** | **4870** | **4025** | **4990** | **4120** | **5150** | **4210** | **6950** | **5650** | **10,210** | **9050** |
| October | 2300 | 1450 | 3210 | 2300 | 3950 | 2460 | 3770 | 2610 | 5940 | 4400 | 8210 | 5570 |
| November | 2010 | 1000 | 2800 | 1750 | 2610 | 1930 | 2840 | 2010 | 4850 | 2990 | 6520 | 3750 |
| December | 1590 | 910 | 2460 | 1410 | 2570 | 1670 | 2700 | 1900 | 4880 | 2810 | 6520 | 3670 |
| January | 1500 | 720 | 2210 | 1340 | 2370 | 1540 | 2710 | 1960 | 4810 | 2760 | 6550 | 3650 |
| **Average (Post-monsoon)** | **1850** | **1020** | **2670** | **1700** | **2875** | **1900** | **3005** | **2120** | **5120** | **3240** | **6950** | **4160** |

(*Source* Sectional office at Ghatal and Tamluk, Irrigation and Waterways Department, West Bengal and Field measurement)

sedimentation. In dry period the scarcity of rainfall reduces the low tide discharge of water, *reduction of stream energy* and transporting capacity. But in monsoon season occurrence of huge rainfall increases the low tide water discharge and stream energy. Because of this *sedimentation rate* is high during dry season but less in monsoon season (Table 1.2).

# References

Allan JD (1995) Stream ecology, the structure and function of running waters. Chapman and Hall, New York

Bathurst JC (1993) Flow resistance through the channel network. In: Beven K, Kirkby MJ (eds) Channel network hydrology. Wiley, Chichester, pp 69–98

Chiu CL, Tung NC (2002) Maximum velocity and regularities in open-channel flow. J Hydraul Eng 128(4):390–398

Dingman SL (1989) Probability distribution of velocity in natural channel cross-sections. Water Resour Res 25:509–528

French RH (1986) Open channel hydraulics. McGraw-Hill, New York

Gorycki MA (1973) Hydraulic drag: a meander initiating mechanism. Geol Soc Amer Bull 84:175–186

Grandgirard V, Legoulven P, Calvez R, Lamouroux N (2013) Velocity and depth distributions in stream reaches: testing European models in Ecuador. J Hydraul Eng-ASCE 139(7):794–798

Hickin EJ (1978) Mean flow structure in meanders of the Squamish River, British Columbia. Canadian Jour Earth Sci 15(11):1833–1849

Hubbell DW, Matejka DQ (1959) Investigations of sediment transportation Middle Loup River at Dunning, Nebraska. US Geol Surv Water Suppl Pap 1476:123

Knighton AD (1979) Comments on long-quadratic relations in hydraulic geometry. Earth Surf Process Landf 4:205–210

Knighton D (1998) Fluvial forms and processes: a new perspective. Arnold, London, p 383

Lamouroux N (1998) Depth probability distributions in stream reaches. J Hydraul Eng 124 (2):224–227

Lamouroux N, Souchon Y, Herouin E (1995) Predicting velocity frequency distributions in stream reaches. Wat Resour Res 31(9):2367–2375

Legleiter CJ, Kyriakidis PC, McDonald RR et al (2011) Effects of uncertain topographic input data on two-dimensional flow modelling in a gravel-bed river. Wat Resour Res 47(3):W03518. doi:10.1029/2010WR009618

Leopold LB, Maddock T (1953) The hydraulic geometry of stream channels and some physiographic implications. US Geol Survey Prof paper, p 252

Leopold LB, Wolman MG, Miller JP (1964) Fluvial processes in geomorphology. Freeman, San Francisco, CA, p 522

Maity SK, Maiti RK (2012) Impact of sedimentation on development and shifting of shoal area, pools and riffles and thalweg position at lower reach of Rupnarayan River-a case study. Indian J Power River Val Dev 24:46–54

Mérigoux S, Lamouroux N, Olivier JM et al (2009) Invertebrate hydraulic preferences and predicted impacts of changes in discharge in a large river. Freshw Biol 54(6):1343–1356

Morisawa M (1985) Rivers: forms and process. Longman Inc, New York

Nepf HM (1999) Drag, turbulence, and diffusion through emergent vegetation. Wat Resour Res 35:479–489

Nordin CF (1963) A preliminary study of sediment transport parameters, Rio Puerco near Bernado. New Mexico, US Geol Survey Prof Paper, p 462

Park CC (1977) World-wide variations in hydraulic geometry exponents of stream channels: an analysis and some observations. J Hydrol 33:133–146

Reynolds AJ (1874) Turbulent flows in engineering. Wiley, Chichester

Richards KS (1973) Hydraulic geometry and channel roughness—a non- linear system. Amer J Sci 273:877–896

Rosenfeld JS, Campbell K, Leung ES et al (2011) Habitat effects on depth and velocity frequency distributions: implications for modeling hydraulic variation and fish habitat suitability in streams. Geomorphology 130(3–4):127–135

Savini J, Bodhaine GL (1971) Analysis of current meter data at Columbia River Gaging Station, Washington and Oregon, US Geol Surv Water Suppl Pap, 1869-F, p 59

Stewardson MJ (1999) Characterizing and modeling the hydraulic environment of streams. Dissertation, University of Melbourne, Victoria

Wolman MG, Gerson R (1978) Relative scales of time and effectiveness of climate in watershed geomorphology. Earth Surf Process Landf 3:189–208

# Chapter 4
# Tidal Character

**Abstract** Being a part of *Hoogly estuary*, the lower reach of the Rupnarayan River is affected by semi-diurnal tide of Bay of Bengal. *Tidal range, tidal prism, tidal asymmetry* and variation of *tidal velocity* in different phases play dominant role to control the mechanism and rate of *sedimentation* in the lower reach. Tidal gauge data have been collected at an interval of one hour at different gauge stations. Tidal range increases from upstream to downstream (2.8 m at Kolaghat, 3.75 m at Soyadighi, 3.85 m at Anantapur, 4.05 m at Pyratungi, 4.3 m at Dhanipur and 4.4 m at Geonkhali). Extensive *intertidal areas* towards downstream receive more water and sediment during high tide than areas towards upstream and enhances the sedimentation rate towards downstream. Mid-tide water velocity is more (1.31–2.15 m/s) in all the places than full-tide water velocity (1.25–2.08 m/s). The *mudflats* and marshy areas which lie above mid-tide level are flooded over 700 times every year during flood tide and getting sedimented. High tide duration is shorter by 2–6 h than that of low tide and this tidal asymmetry results swifter flow with greater energy during high tide leading to landward transport of sediment. The sluggish low tide discharge over longer duration (8–9 h) allows sufficient opportunity for settlement of sediments.

**Keywords** Estuary · Tidal range · Tidal asymmetry · Tidal velocity · Mudflats · Sedimentation

## 4.1 Introduction

The volume water exchanged during a tidal cycle through flood and ebb tides helps to classify estuaries into flood dominant, ebb dominant, or co dominant, where the flood and ebb are of equal magnitude. Being a part of *Hoogly estuary*, the lower reach of the Rupnarayan River is affected by semi-diurnal tide of high *tidal asymmetry*. Leopold et al. (1964) mentioned that the discharge of a river is not dependent on the size of the channel, but estuary discharge is dependent on channel size. Estuaries receive an amount of water not in proportion to a fixed drainage area but determined by the capacity of the channel supplying the water. According to

S. Kumar Maity and R. Maiti, *Sedimentation in the Rupnarayan River*,
SpringerBriefs in Earth Sciences, https://doi.org/10.1007/978-3-319-62304-7_4

Wright et al. (1973), flow and channel of an estuary have developed together and simultaneous co-adjustment of both process and form has yielded an equilibrium situation. The two-way flow of estuaries, the current set up by the mixing of fresh and saline water and the continuous variations which take place in both velocity and discharge through the tidal cycle provide a marked contrast with fluvial processes and increase the rate of sedimentation dramatically (Pethick 1984). French (1997) mentioned that in estuaries, currents reach maximum landward velocity at *mid-flood tide*, decrease to zero at high and low tide, before reversing to maximum ebb velocity midway through the ebb. If the tidal velocity becomes zero, finest sediments can settle out and becomes deposited due to *flocculation* of particles helped by electrolytic imbalances due to mixing of fresh and saline water. According to Rhodes (1950) as the salt water wedge approaches towards upstream, the river flow mixes with salt water and loses its velocity and the bed load will cease to move and coarse fraction of the suspended load will settle down following *Stoke's law*. In case of tidal asymmetry, during shorter high tide, the fixed volume of water moves faster than on the longer low tide. As a result the velocity and stream energy (sediment transport capacity) in high tide is more than in low tide condition (French 1997). The upper part of estuarine channels is predominantly *depositional environments* in which the trapped sediments are laid down and shaped by the tidal and fresh water flows (Dyer 1973).

## 4.2  Data Collection and Methods

Water level data during high and low tide have been measured at an interval of one hour at different *gauge stations* (6 stations) (Fig. 4.1a, b) along the lower reach of the Rupnarayan River. Tidal gauge data have been collected from Kolaghat Thermal Power Station (K.T.P.S.) and the Sectional Office at Ghatal and Tamluk, under the Irrigation and Waterways Department of Government of West Bengal respectively. Hourly tidal data for the monsoon period have also been collected for some of the stations (Bandar, Ranichak and Gopigaung), at middle and the upper

**Fig. 4.1**  Tidal gauge station at Kolaghat (**a**) and Geonkhali (**b**)

reach of the river for proper understanding of the modifications of tide during its journey and its impact. *Tidal range* is calculated by differentiating the water level during extreme high tide and extreme low tide within a day. *Tidal asymmetry* is understood by measuring the duration of high tide and low tide at a particular station. *Tidal prism* (volume of water entering in a tidal region during high tide) is calculated by measuring the surface area of the water at high tide, and multiplying it by the tidal range (following Carter 1988).

## 4.3 Tidal Velocity

In a fluvial channel the fastest current velocities are associated with the highest water levels in the channel such as those attained during flood conditions. Consequently maximum discharge occurs when the channel is filled and thus most work is done in a fluvial channel at *bankful stage*, which is therefore sometimes called dominant discharge (Pethick 1984). But in estuaries the pattern of velocity, water level and discharge is more complicated. In most estuaries the partial reflection of the tidal wave means that maximum velocities are attained at mid-tide when the channel is only half-full (Sverdup et al. 1942). Arguing from the analogy of the fluvial channel it could be said that this mid-tide section of the estuary is the hydraulically important channel in which dominant discharge is attained. All the places at lower reach of the Rupnarayan River attain maximum velocity during mid-tide condition than during full-tide condition (Table 4.1). At Geonkhali average velocity during mid-tide (2.15 m/s) is more than the average velocity during full-tide (2.08 m/s). *Mid-tide velocity* (1.75 m/s) is more than full-tide velocity (1.68 m/s) at Dhanipur. At Soyadighi the average velocity during mid-tide is 1.44 m/s but during full-tide it is 1.40 m/s. The average velocity during mid-tide and full-tide are 1.75 m/s and 1.68 m/s respectively at Kolaghat (Table 4.1). Consequently the *mudflats* and marshes which lie above mid-tide are analogous to the *flood-plain* of a river. River flood-plain is covered by water only once in two or three years on an average, but the mudflats of the estuary are flooded over 700 times every year. So, *sedimentation* rate is high on estuarine mudflats. Similarly the central channel below mid-tide will attain dominant discharges twice a day and will be swept clear of sediments (Perkins 1974).

**Table 4.1** Water velocity during mid-tide and full-tide

| Reach | Geonkhali | Dhanipur | Pyratungi | Anantapur | Soyadighi | Kolaghat |
|---|---|---|---|---|---|---|
| Average velocity in mid-tide (m/sec) | 2.15 | 1.75 | 1.55 | 1.68 | 1.44 | 1.75 |
| Average velocity in full tide (m/sec) | 2.08 | 1.68 | 1.50 | 1.62 | 1.40 | 1.68 |

(*Source* Field measurement)

## 4.4   Tidal Range

Tidal range is the difference in elevation between high and low water marks. Hayes (1975) and Davies (1964) has divided the *tidal range* into three classes (i) *Micro-tidal (<2 m)* (ii) *Meso-tidal (2–4 m)* (iii) *Macro-tidal (>4 m)*. In Rupnarayan River the tidal range continues to decrease from downstream (Geonkhali) to upstream (Bandar) (Table 4.2 and Fig. 4.2). Based on the tidal range, Pyratungi, Dhanipur and Geonkhali reach are under Macro-tidal (>4 m) range, whereas Kolaghat (Dainan) and Soyadighi and Anantapur reach are under Meso-tidal (2–4 m) range. In the middle and upper course of the Rupnarayan River Gopigaung, Ranichak and Bandar reach are under Micro-Tidal range (<2 m) (Table 4.2 and Fig. 4.2). Tidal gauge level data collected from Sectional Office at Ghatal and Tamluk, Irrigation and Waterways Department, West Bengal indicates that tidal range is 4.4 m, 4.3 m, 4.05 m, 3.85 and 3.75 m at Geonkhali, Dhanipur, Pyratungi, Anantapur and Soyadighi respectively. It has decreased to 2.8 m near Dainan (Kolaghat) (Table 4.2 and Fig. 4.2).

As the wave moves up-estuary, *frictional effects* cause the wave's energy to be dissipated and the wave height therefore decreases since wave energy is proportional to wave height (Pethick 1984). In most of the cases this energy dissipation means that as the tides moves landward, it gradually flattens (decrease of tidal range)  and ultimately disappears. In other cases however some energy may be reflected from the banks of the estuary channels and a reflected wave will move back seawards against the advancing tide (Pethick 1984).

### 4.4.1   Tidal Range and Sedimentation

*Tidal range* is an important determinant to control the spatial variation sedimentation rate at lower reach of the Rupnarayan River. The extension of intertidal areas that receive water during high tide depends on tidal range (Pethick 1984). Extensive *intertidal areas* towards downstream receive more water and sediment during high tide than areas towards upstream. That's why the average sedimentation rate is high towards downstream than that in upstream. Average rate of sedimentation is 2.69 mm/h at Geonkhali and 3.23 mm/h at Dhanipur. But it decreases to

**Table 4.2**  Tidal range and tidal asymmetry at lower reach of the Rupnarayan River

| Reach | Geonkhali | Dhanipur | Pyratungi | Anantapur | Soyadighi | Kolaghat |
|---|---|---|---|---|---|---|
| Tidal Range (m) | 4.4 | 4.3 | 4.05 | 3.85 | 3.75 | 2.8 |
| Tidal asymmetry | 2 h | 3 h | 3 h 30 min | 3 h 45 min | 4 h 30 min | 6 h |

(*Source* Sectional Office at Ghatal and Tamluk, Irrigation and Waterways Department, West Bengal and Field measurement)

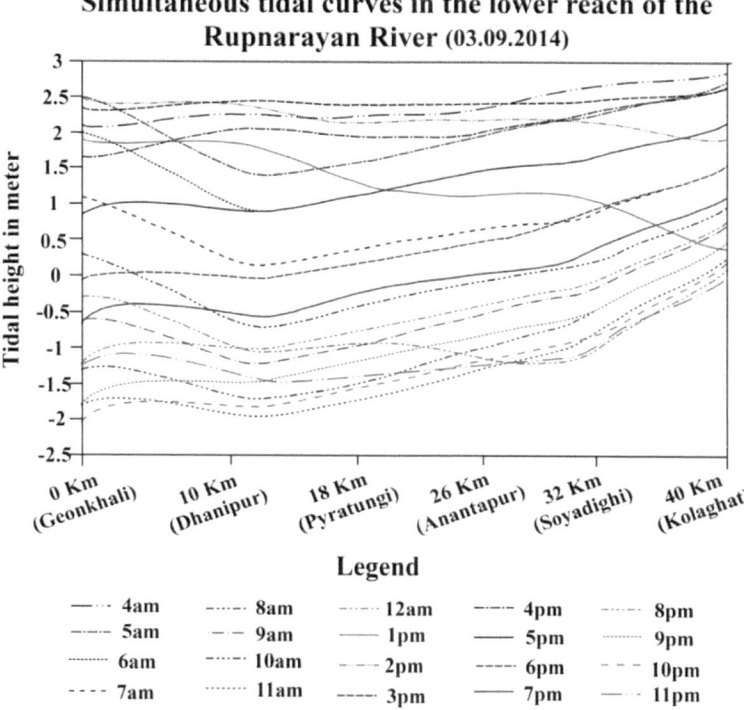

**Fig. 4.2** Magnitude of tidal range at lower reach of the Rupnarayan River (Maity and Maiti 2016). (*Source* Sectional Office at Ghatal and Tamluk, Irrigation and Waterways Department, West Bengal)

2.29 mm/h at Soyadighi and 2.47 mm/h at Kolaghat (Table 1.2). Though, at the lower reach the sedimentation rate is haphazard but it continuously decreases towards upstream and becomes insignificant and negligible at Bandar.

## 4.5 Tidal Asymmetry

*Tidal asymmetry* means the unequal length of flood and ebb period in a tidal region. If the tidal wave moves up the estuary there is a gradual increase of tidal asymmetry (Pethick 1984). The leading edge of the wave (flood tide) becomes steeper while the trailing edge (ebb tide) becomes flatter. This asymmetry is noted at lower reach of the Rupnarayan River, which leads to extremely important variations in estuarine tidal processes. In this area, the leading edge covers 3–4 h and the trailing edge covers 8–9 h. Tidal asymmetry is 2 h at Geonkhali reach and it increases to 3 h at Dhanipur reach. Again it increases to 3 h 30 min at Pyratungi, 3 h 45 min at Anantapur, 4 h 30 min at Soyadighi and finally becomes 6 h at Kolaghat reach

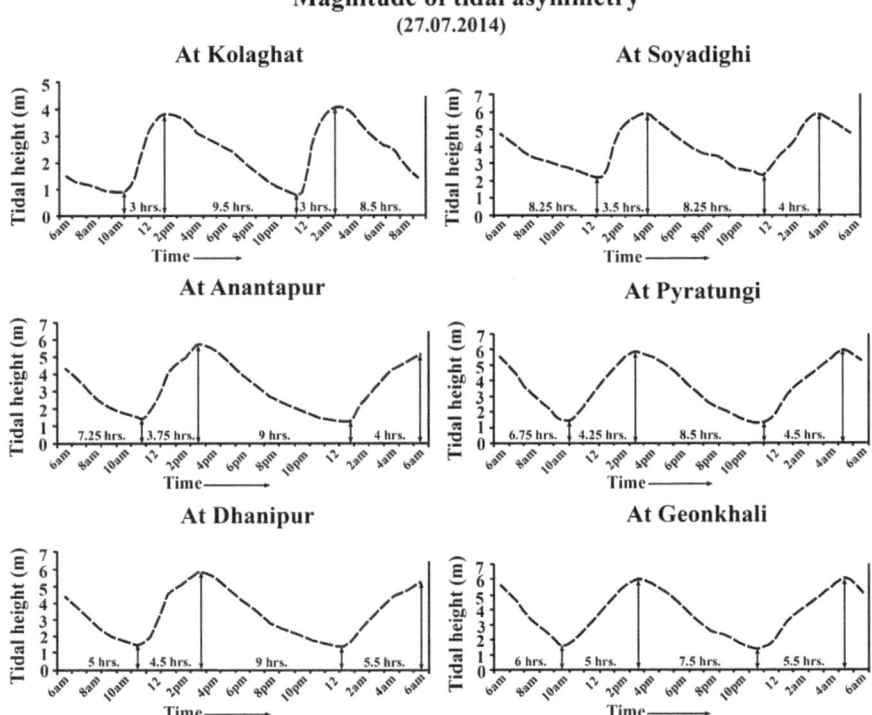

**Fig. 4.3** Degree of tidal asymmetry at different places in the lower reach. (*Source* Sectional office at Ghatal and Tamluk, Irrigation and Waterways Department, West Bengal and Field measurement)

(Table 4.2 and Fig. 4.3). From the measurement of Kolaghat Thermal Power Project (Fig. 4.4) it is represented that tidal asymmetry is 2 h 15 min at Geonkhali, 3 h 30 min at Dhanipur, 4 h at Pyratungi and 4 h 30 min at Soyadighi respectively. The asymmetry has increased to 6 h 15 min at Dainan (Kolaghat). So, tidal asymmetry increases upstream. Difference in the wave-form-velocity at crest and trough of the tidal wave is the main causal mechanism of tidal asymmetry. Crest and trough are moving in appreciably different water depths since the height of the tidal wave itself must be considered as well as the still- water depth. If 'g' is gravitational acceleration, 'D' is channel depth and 'H' is tidal wave amplitude then the velocity of the crest is $C_{crest} = \sqrt{g\left(D + \frac{1}{2}H\right)}$, while the velocity of the trough is $C_{trough} = \sqrt{g\left(D - \frac{1}{2}H\right)}$. As a result the crest (high tide wave) overtakes the trough (low tide wave) so that the tidal wave becomes increasingly asymmetric (Pethick 1984).

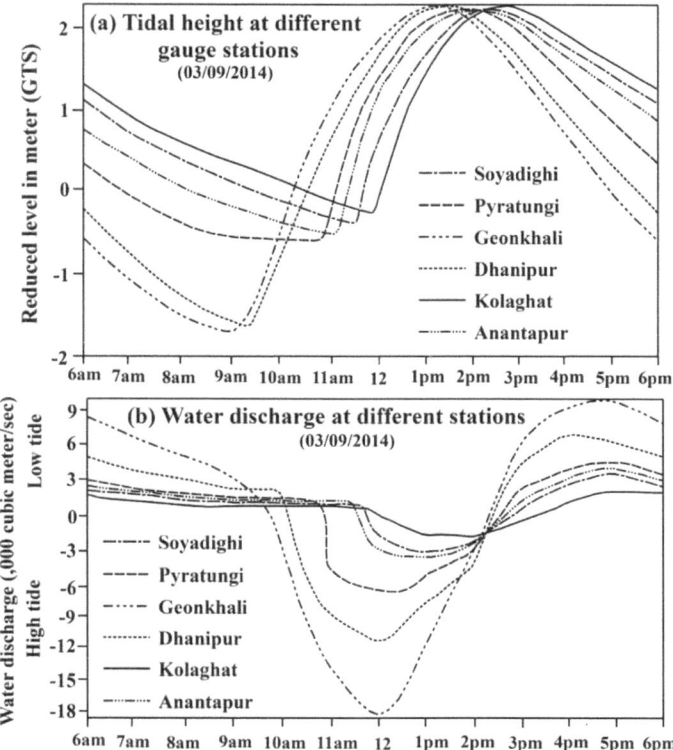

**Fig. 4.4** Gauge height (**a**) and water discharge (**b**) at different stations (Maity and Maiti 2016). (*Source* Kolaghat Thermal Power Project)

### 4.5.1  Tidal Asymmetry, Water Velocity, Stream Energy and Sedimentation

*Tidal asymmetry* is the most important cause for increasing sedimentation at lower reach of the Rupnarayan River. During high tide, water and sediment moves faster within short time but in low tide, same amount of water and sediment is discharged for longer duration. As a result the velocity, energy and sediment transport capacity is more in high tide than in low tide. The reduction of velocity and stream energy in low tide condition leads to sedimentation. Hence, the region (upper part of the Hoogly estuary) will experience a net input and *trapping of sediment*. That's why the estuaries are predominantly depositional environment in which the trapped sediments are laid down and shaped by the tidal and fresh water flows (Dyer 1973).

## 4.6   Tidal Prism

The volume of water that penetrates in a tidal region during high tide is called *Tidal prism*. Jarrett (1976) expressed the relationship between tidal prism (P) and the cross-sectional area of a tidal region. The volume of water (tidal prism) can be estimated by measuring the surface area of the water at high tide, and multiplying it by the tidal range (Carter 1988). The measurement of the authority of Kolaghat Thermal Power Project in the year 2014 reveals that, at the lower reach huge volume of water enters towards upstream areas during high tide condition which allows the penetration of voluminous amount of sediment towards upstream (Fig. 4.4). During monsoon season, due to occurrence of huge rainfall the upward and downward movement (in high tide and low tide respectively) of water is almost equal giving equal opportunity of sediment transport in both the directions (Maity and Maiti 2016). That's why the sedimentation rate is reduced in this season (Table 1.2). But in dry season (pre-monsoon and post-monsoon) low tide the discharge of water from upstream is less compared to penetration of water towards upstream during high tide. So, there is the possibility of penetrating more sediment upstream than being drained towards downstream and the rate of sedimentation is accelerated (Table 1.2).

### 4.6.1   Tidal Prism and Sediment Transport Power

In case of a tidal river the mechanism of *sediment deposition* is the combined effect of the interaction between terrestrial discharge from upstream and the inward penetration of water during high tide. The riverine discharge is impounded during high tide and released during ebb tide. The sediments which are transported by the littoral drift towards downstream reaches the river mouth and are picked up by the *tidal water* and again redistributed towards upstream by the flood current (Defant 1961). The regular influxes of tidal water through the river Rupnarayan visit this area twice a day. Huge volumes of sediments are being transported towards upstream during high tide and are deposited on the bed of Rupnarayan River in the absence of sufficient upland discharge in dry seasons. During rainy season (June, July, August and September) the influence of high tide is reduced due to the voluminous *terrestrial discharge* from the upper catchment of the Rupnarayan River. Inspite of this direct impact of the tidal phenomena, the inoculation of saline water during high tide into the sweet water play an important role to control mechanism of sediment deposition. As the saline water approaches inward from mouth ward side of the estuary, the fresh riverine water flow interact with the salt water wedge and loses its velocity. The sediments which were transported as bed load will cease to move and the coarser fractions of the suspended load will also be deposited according to *Stoke's law*. Comparatively finer and colloidal suspended sediment will settle down very slowly, whereas the sediments which are subjected to electro-chemical reaction will flocculate after meeting with the saline water (Rhodes 1950).

Tidal current is generally non-steady and non-uniform in nature and sediments are transported in different ways including bed load, suspended lad or colloidal load in a river. Because of this, a correct expression of sediment transportation power of a tidal current is very difficult to establish. Generally, the transportation power of a stream is directly proportional to the square, cube or any higher power of mean velocity of water flow ($V^m$ with $m > 1$) and the mean water velocity is again proportional to the discharge of water ($Q$). So, the transportation power of the stream will then be proportional to $Q^m$, m being greater than unity. The discharge of water and the corresponding sediment transportation power ($Q^m$) at any point can be plotted against time over a complete tidal period. In the lower reach of the Rupnarayan River the high tide and the low tide duration ranges between 3–5 h and 7–9 h (Fig. 4.5) respectively. Figure 4.5 represents that area under the Q curve is the total volume of water discharge while the area under the $Q^m$ curve is the work done in transporting sediment within the same time. The difference in the area enclosed by the $Q^m$ curve for the high and low tide indicates the net volume of sediment which has been transported either towards upstream or downstream as the case may be. When the tidal wave having a certain discharge of $Q_0$, moves towards upstream in the river Rupnarayan, the symmetrical water discharge ($Q$) curve experience some degree of deformation and will be asymmetric in shape. It is because of the fact that the crest of every tidal wave partially overtakes the trough preceding it. As an outcome of this, the Q curve at a little distance towards upstream shows a sharper rise in water discharge with a corresponding shorter time period for the high tide and a sluggish outflow and consequently a longer time for the low tide. In this condition the area under the $Q^m$ curve for the high tide is usually greater than that of the low tide. The net result is the excess transportation of sediment towards upstream which indicates the deposition of sediment on the river bed (Fig. 4.5) (Maity and Maiti 2016).

But during rainy season (June to September) this condition becomes reverse due to voluminous riverine discharge from the upland area. The additional upland discharge during freshet leads to increase of stream velocity and good amount of washing out of sediment from the river bed. In dry season, the sediments supplied from the tributaries of Rupnarayan River initially move downward and then get deposited on the river bed causing sporadic shoaling in the lower reach of the river. But in freshet as the low tide becomes stronger due to the *voluminous upland discharge*, it enhances the seaward movement of sediments and restricts the deposition of sediments, causing the scouring on the river bed. As a result the best possible depth of the river is attained and the net result can be represented by the difference between the areas enclosed by the $Q^m$ curve of the high tide and the $(Q + Q_0)^m$ curve of the low tide as shown in Fig. 4.5.

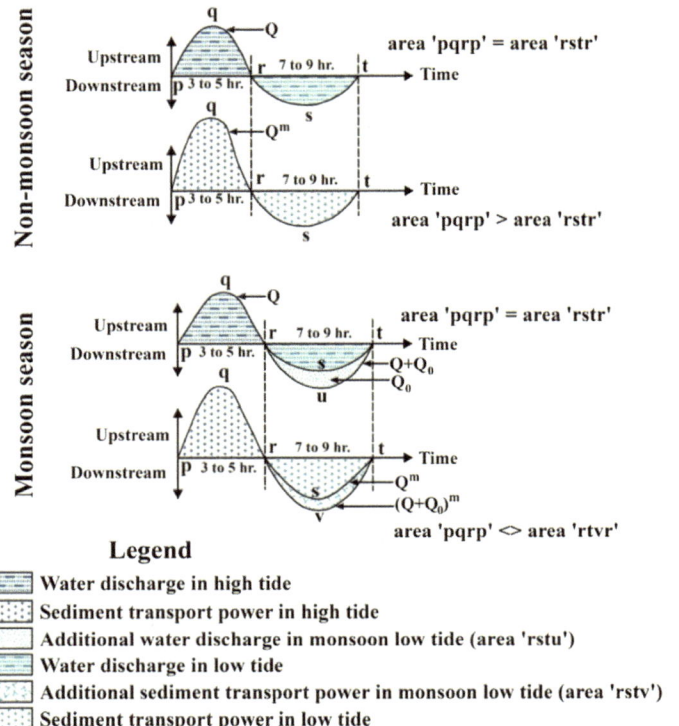

Fig. 4.5 Relation between Tidal prism and sediment transport capacity (Maity and Maiti 2016)

# References

Carter RWG (1988) Coastal environment: an introduction to the physical, ecological and cultural system of coastlines. Academic, London

Davies JL (1964) A morphogenic approach to the worlds' shorelines. Z Geomorph 8:127–142

Defant A (1961) Physical oceanography. Pergamon, New York

Dyer K (1973) Estuaries: a physical introduction. Wiley, London

French WP (1997) Coastal and estuarine management. Routledge environmental management series. Routledge, London, pp 1–219

Hayes MO (1975) Morphology of sand accumulation in estuaries. In: Cronin L (ed) Estuarine research, vol II. Academic, New York

Jarrett JT (1976) Tidal prism-inlet area relationships. GITI Report 3, U.S. Army Engineering Waterways Experiment Station, Vicksburg, Mississippi, USA

Leopold LB, Wolman MG, Miller JP (1964) Fluvial processes in geomorphology. Freeman, San Francisco, CA, p 522

Maity SK, Maiti RK (2016) Tidal impact leading to sedimentation at lower reach of the Rupnarayan River, West Bengal, India. Indian Jour Mar Sci 45(10):1349–1356

Perkins FJ (1974) The biology of estuaries and coastal waters. Academic, London

Pethick JS (1984) An introduction to coastal geomorphology. Arnold, London

Rhodes RF (1950) Effects of salinity on current velocities. US Corps of Engineers, Committees Tidal Hydraulics, report No 1, p 94

Sverdup HV, Johnson MW, Fleming RH (1942) The Oceans. Prentice Hall, New York

Wright LD, Coleman JM, Thom BG (1973) Processes of channel development in a high tide range environment: Cambridge Gulf-Ord River Delta. J Geol 81:15–41

# Chapter 5
# Sediment Load: Concentration and Transport

**Abstract** Amount of *suspended sediment* in water is measured by collecting water samples from different depths during *high and low tide*. Rate of transportation of *bed load* has been computed using well accepted empirical equations. Suspended sediment amount ranges between 3.1 and 5.05 gm/l in non-monsoon season and 4.97 and 6.5 gm/l in monsoon season. During non-monsoon period, upstream penetration of suspended sediment is more during high tide ($3.71 \times 10^7$ to $1.29 \times 10^8$ metric tons/year) than that is discharged towards downstream during low tide ($2.5 \times 10^7$ to $1.0 \times 10^8$ metric tons/year) and accelerates the rate of *sedimentation*. But in monsoon, the transport of suspended sediment during high tide ($7.5 \times 10^7$ to $2.45 \times 10^8$ metric tons/year) and low tide ($7.3 \times 10^7$ to $2.3 \times 10^8$ metric tons/year) is almost equal which restricts the *sedimentation* rate. Rate of suspended sediment transport is strongly affected by variation of water discharge ($r = 0.908$) and moderately affected by sediment concentration ($r = 0.553$). Increasing tendency of water velocity and discharge towards downstream leads to the increase of bed load transport rate from Kolaghat to Geonkhali. It varies from 0.1905 to 6.52985 kg/m/sec, 0.5008 to 14.74893 kg/m/sec and 0.2318 to 6.31764 kg/m/sec in pre-monsoon, monsoon and post-monsoon season respectively. In non-monsoon season, the transport of bed load is more during high tide than during low tide, but the transport of bed load is almost equal in both the tidal phases in monsoon season.

**Keywords** Sediment load · Suspended load · Bed load · High and low tide · Sedimentation

## 5.1 Introduction

*Sediment discharge* from the river and its associated material fluxes are the major processes of material transfer from land to the ocean (Allan 1986; Walling 1989; Meybeck 1993, 1999). The discharge of sediment is linked in a complex manner, to large numbers of flow and sediment parameters like, the depth, width, density, energy gradient, temperature, viscosity and water turbulence, along with the shape,

© The Author(s) 2018
S. Kumar Maity and R. Maiti, *Sedimentation in the Rupnarayan River*,
SpringerBriefs in Earth Sciences, https://doi.org/10.1007/978-3-319-62304-7_5

size, concentration, density and cohesiveness of the moving sediment particles (Milliman and Meade 1983). Egiazaroff (1965), Ashida and Michiue (1971), Hayashi and Ichibashi (1980), Parker et al. (1982), Andrews (1983), Proffit and Sutherland (1983), Bridge and Bennett (1992), Fang and Yu (1998) and Karim (1998) stated that in case of non-uniform sediment, the coarser particles on the river bed are easily entrained than the uniform sediment of equivalent sizes, because they have more possibility of exposure to the water flow. But the situation is different for the finer sediment particles of the river bed because the finer particles are sheltered by the coarser fractions. Therefore, it is required to think about the hiding and exposure effect in case modeling of non-uniform sediment transport. Hsu and Holly (1992) proposed a technique for the prediction of the gradation of non-uniform bed-load by considering the probability and availability of the moving sediment particles. Empirical functions were established by Samaga et al. (1986b) and Karim (1998) to estimate the transportation rates of suspended load and bed load. Fournier (1960) identified the relationship between climatic type and amount of *sediment yield*. Gilbert (1914) correlated bed load transport in terms of discharge, slope and sediment size. Einstein (1942) took a stochastic approach in bed load transport by relating movement of grains to random fluctuations in velocity. Colby (1957) stated that most of the variables influencing total sediment transport are related to mean stream velocity. Leopold and Wolman (1953) determined the relationship between load and discharge. They mentioned that at a station the suspended load increases more rapidly than the water discharge and a narrow, deep channel is most efficient for carrying a high suspended load. Wilcock and McArdell (1993) defined a region of partial sediment transport, in which only some of the sediment particles on the river bed are movable at a particular time even though all sediments may eventually take part in the motion. In the present work the amount of bed load transport has been estimated using *empirical equations* of different researchers and suspended load is measured by collecting water samples from different depths of the river under study. The spatial and seasonal variation of sediment load is clarified in relation to the fluctuation of water discharge and stream energy.

## 5.2  Field Monitoring and Applied Methodology

All the required components and variables (like river depth, water velocity, water discharge, sediment concentration, river bed slope etc.) have been monitored and measured simultaneously during high and low tide in three seasons (Pre-monsoon, Monsoon and Post-monsoon) at six stations (Kolaghat, Soyadighi, Anantapur, Pyratungi, Dhanipur and Geonkhali) to be fixed all along the lower reach of the Rupnarayan River (Fig. 1.1e). Amount of *suspended sediment* in water is measured by collecting the water samples from different depth during high and low tide for twelve months of the year dividing into three seasons. Suspended sediment transport is computed following equation 5.1 (Hickin 1995)

$$Qs = CsQ \qquad (5.1)$$

Where, $Qs$ = suspended sediment discharge, $Cs$ = suspended sediment concentration ($kg/m^3$), $Q$ = the discharge of water ($m^3/sec$).

For example, if the average annual suspended sediment load in a river is 4 gm/l ($4000 \ gm/m^3 = 4 \ kg/m^3$) and the average annual discharge is 5000 $m^3/sec$, then the amount of suspended sediment transport ($Qs$) will be-

$Qs = 4 \ kg/m^3 \times 5000 \ m^3/sec = 20000 \ kg/sec = 12 \times 10^5 \ kg/minute = 72 \times 10^6 \ kg/h = 1728 \times 10^6 \ kg/day = 6.3072 \times 10^{11} \ kg/day = 6.3072 \times 10^8$ metric tons/year.

In case of my study the duration of high tide and low tide (hour) have been taken into consideration for the calculation of sediment discharge per day in place of 24 h.

Empirical equations of different researchers have been used to estimate the amount of *bed load transport-*

$$q_b^{\frac{2}{3}} = 250q^{\frac{2}{3}}S - 42.5d_s \qquad (5.2) \qquad \text{(Meyer-Peter-Muller 1948)}$$

$q_b$ = unit bed load transport rate (kg/m/sec), $q$ = unit water discharge (cubic meter/sec), $d_s$ = diameter of bed material in meter, $S$ = slope of the channel bed.

$$qb = \left(\frac{2}{3}\right)(\rho s - \rho)gd_s \left(\frac{L}{T}\right)p \qquad (5.3) \qquad \text{(Einstein 1950)}$$

$\rho s$ = sediment density, $\rho$ = water density, $g$ = gravitational acceleration, $d_s$ = diameter of sediment grain, $L$ = jump length, $T$ = time consumed for erosion, $p$ = the probability of particle being eroded from river bed.

$$q^* = 0.635r\sqrt{\tau^*}\left[1 - \frac{1}{\sigma r}\ln(1 + \sigma r)\right] \qquad (5.4) \qquad \text{(Yalin 1963)}$$

$$r = \frac{\tau^*}{\tau^*_{crit}} - 1 \qquad (5.4a)$$

$$\sigma = 2.45\frac{\sqrt{\tau^*_{crit}}}{(\rho_s/\rho)^{0.4}} \qquad (5.4b)$$

$q^*$ = unit bed load transport rate (kg/m/sec), $\tau^*$ = available shear stress, $\tau^*_{crit}$ = critical shear stress, $\rho s$ = sediment density, $\rho$ = water density.

$$q'b = m'_b g U_b \qquad\qquad (5.5) \quad \text{(Bagnold 1966)}$$

$m'_b g$ = normal stress, $U_b$ = mean transport velocity of bed load particle.

$$Q_T = 0.1 \left[ \gamma_s \left( \frac{\gamma_s - \gamma}{\gamma} \right) g d^3 \right]^{\frac{1}{2}} \theta^{\frac{5}{2}} \frac{\theta^{\frac{5}{2}}}{f^1} \qquad\qquad (5.6) \quad \text{(Engelund-Hansen 1967)}$$

$Q_T$ = bed load transport rate, $\gamma_s$ = specific weight of sediment, $\gamma$ = specific weight of water, $g$ = gravitational acceleration, $d$ = diameter of bed particle, $\theta$ = dimensionless shear stress parameter, $f^1$ = friction factor.

$$\log C_t = I + J \log \left( \frac{VS}{\omega} - \frac{V_c S}{\omega} \right) \qquad\qquad (5.7) \quad \text{(Yang 1973)}$$

$$I = 5.435 - 0.286 \log \left( \frac{\omega d_s}{v} \right) - 0.457 \log \left( \frac{u_*}{\omega} \right) \qquad\qquad (5.7a)$$

$$J = 1.799 - 0.409 \log \left( \frac{\omega d_s}{v} \right) - 0.314 \log \left( \frac{u_*}{\omega} \right) \qquad\qquad (5.7b)$$

$C_t$ = total bed material transport, $VS$ = unit stream power, $V_c S$ = critical unit stream power at incipient motion, $\omega$ = terminal fall velocity of sediment particle, $d_s$ = particle diameter, $v$ = kinematic viscosity of flow, $u_*$ = shear velocity

$$Q_T = Q.X \qquad\qquad (5.8) \quad \text{(Ackers-White 1973)}$$

$$X = G_{gr} \frac{d}{h} \frac{\gamma_s}{\gamma} \left( \frac{v}{U_*} \right)^{C_1} \qquad\qquad (5.8a)$$

$Q_T$ = bed load transport rate, $Q$ = water discharge, $X$ = sediment concentration, $G_{gr}$ = dimensionless sediment transport function, $d$ = average particle diameter, $h$ = water depth, $\gamma_s$ = specific weight of sediment, $\gamma$ = specific weight of water, $v$ = average flow velocity, $U_*$ = shear velocity, $C_1 = 1.00 - 0.56 \log d_*$, $d_*$ = dimensionless particle diameter.

$$qb = C_b U_b \delta_b \qquad\qquad (5.9) \quad \text{(Van Rijn 1984)}$$

$q_b$ = unit bed load transport rate (kg/m/sec), $C_b$ = average bed load concentration, $U_b$ = bed layer particle velocity, $\delta_b$ = saltation height/bed layer thickness.

$$\emptyset_b = 8.5(\tau_*)^{1.8}/[1 + 5.95 \times 10^{-6}(\tau_*)^{-4.7}]^{1.45} \qquad (5.10) \qquad \text{(Samaga et al 1986a)}$$

$\emptyset_{b=}$ bed lad transport rate in mass per unit width, $\tau_{*=}$ shear stress parameter similar to Shields' parameter with minor modifications.

## 5.3  Amount of Suspended Sediment in Water

The clastic (particulate) material (<0.062 mm in diameter) which moves through the channel within the water column is considered as suspended sediment. The upward flux of turbulence generated at the bed of the river channel keeps the materials, mainly silt and sand in suspension (Sundborg 1956). Amount of suspended sediment in water is an important component of total sediment transport in a river. It controls the *morphology of the channel* by influencing channel resistance, turbulence and stream velocity (Colby 1964). It is again affected by watershed characteristics (geology, land-use and climate etc.) and hydraulic characteristics of the stream (depth, width etc.) (Bogardi 1974). Because of this the suspended sediment concentration in river water varies spatially and temporally. In the study area, the suspended sediment concentration in water is quantified by collecting water samples at different depths during low tide and high tide conditions.

### 5.3.1  Fluctuation of Suspended Sediment Amount in Water

Amount of suspended sediment in water varies spatially and seasonally in the lower reach of the Rupnarayan River. Generally, in all the months of dry season, the suspended sediment concentration is always higher in high tide water than in low tide water (Table 5.1). But in the months of monsoon season voluminous supply of water and sediment from upland area increases the concentration of suspended load in low tide water (Table 5.1). During high tide, highest (6.3 gm/l) and lowest (2.8 gm/l) amount of suspended sediment load is measured in the months of July (monsoon) and March (pre-monsoon) at Dhanipur and Geonkhali respectively. In low tide condition, the highest (6.6 gm/l) and lowest (2.8 gm/l) amount of suspended sediment load is measured in the months of August (monsoon) and March (pre-monsoon) at Dhanipur and Pyratungi respectively (Table 5.1). Average highest concentration of suspended sediment is measured at Dhanipur (3.5–6.5 gm/l) in all the seasons. It is also high (3.22–5.95 gm/l) near Kolaghat region compared to other places in the lower reach (Table 5.1). Average highest amount of suspended sediment load at Soyadighi, Anantapur, Pyratungi and Geonkhali are measured as 5.6, 5.85, 5.9 and 5.15gm/l respectively. In pre-monsoon season, average suspended sediment load varies from 3.1 to 4.4 gm/l, in monsoon it varies from 4.97 to 6.5 gm/l and in post-monsoon season it varies from 3.45 to 5.05 gm/l (Table 5.1). Sheppard

**Table 5.1** Amount of suspended sediment in water

| Month | Amount of suspended sediment (gm/liter) | | | | | | | | | | | |
|---|---|---|---|---|---|---|---|---|---|---|---|---|
| | Kolaghat | | Soyadighi | | Anantapur | | Pyratungi | | Dhanipur | | Geonkhali | |
| | High tide | Low tide | High tide | Low tide | High tide | Low tide | High tide | Low tide | High tide | Low tide | High tide | Low tide |
| February | 3.9 | 3.3 | 4.2 | 3.7 | 3.8 | 4.0 | 3.7 | 2.9 | 4.4 | 3.3 | 3.4 | 3.1 |
| March | 4.1 | 3.3 | 3.4 | 3.8 | 3.9 | 3.1 | 4.5 | 2.8 | 4.4 | 3.5 | 2.8 | 2.9 |
| April | 4.2 | 3.1 | 3.9 | 3.3 | 3.3 | 3.1 | 4.1 | 3.4 | 3.9 | 3.6 | 2.9 | 3.3 |
| May | 4.2 | 3.2 | 3.9 | 3.4 | 3.4 | 3.4 | 4.1 | 3.7 | 4.9 | 3.6 | 3.3 | 3.1 |
| **Average (Pre-monsoon)** | **4.1** | **3.22** | **3.85** | **3.55** | **3.6** | **3.4** | **4.1** | **3.2** | **4.4** | **3.5** | **3.11** | **3.1** |
| June | 5.7 | 5.6 | 4.9 | 5.8 | 5.5 | 5.9 | 6.1 | 5.8 | 6.1 | 6.5 | 5.3 | 5.0 |
| July | 6.2 | 5.6 | 5.2 | 5.6 | 5.7 | 5.9 | 6.2 | 5.7 | 6.3 | 6.3 | 5.1 | 5.4 |
| August | 6.1 | 5.6 | 5.3 | 5.6 | 5.7 | 5.8 | 5.7 | 5.5 | 6.2 | 6.6 | 5.3 | 5.1 |
| September | 5.8 | 5.5 | 4.5 | 5.4 | 5.5 | 5.8 | 5.6 | 5.4 | 6.0 | 6.6 | 4.7 | 5.1 |
| **Average (Monsoon)** | **5.95** | **5.58** | **4.97** | **5.6** | **5.6** | **5.85** | **5.9** | **5.6** | **6.15** | **6.5** | **5.1** | **5.15** |
| October | 5.1 | 4.4 | 3.9 | 3.9 | 4.3 | 4.0 | 5.1 | 4.0 | 5.2 | 4.3 | 3.6 | 3.4 |
| November | 4.9 | 4.0 | 4.1 | 3.7 | 4.1 | 4.1 | 5.1 | 4.0 | 5.1 | 4.3 | 3.5 | 3.6 |
| December | 4.8 | 3.7 | 3.9 | 3.6 | 4.1 | 3.6 | 4.5 | 4.3 | 5.0 | 4.1 | 3.3 | 3.5 |
| January | 4.8 | 4.6 | 3.7 | 3.4 | 3.9 | 3.7 | 4.9 | 4.4 | 4.9 | 3.7 | 3.8 | 3.3 |
| **Average (Post-monsoon)** | **4.9** | **4.18** | **3.9** | **3.65** | **4.1** | **3.85** | **4.9** | **4.18** | **5.05** | **4.1** | **3.55** | **3.45** |

(*Source* Field measurement)

(1965) mentioned that generally the concentration of suspended load is expected to decrease towards downstream. But in the study area, having *estuarine characteristics* the mixing of water and sediment during high and low tide leads to an *unsystematic and chaotic* spatial distribution of suspended sediment (Dyer 1973).

## 5.4  Transportation of Suspended Load

In the study area the transport of suspended sediment varies spatially and temporally depending on the volume of water discharge and amount of suspended sediment in water.

### 5.4.1  Fluctuation of Suspended Load Transport

Though few exceptions are found but in general, the average amount of *suspended sediment transport* increases towards downstream (from Kolaghat to Geonkhali) in the lower reach (Table 5.2 and Fig. 5.1). Due to voluminous supply of sediments from upstream region in monsoon season, the *suspended sediment transport* is more in all the months of this season (in both the tidal phases) than pre-monsoon and post-monsoon seasons. During high tide, highest ($2.75 \times 10^8$ metric tons/year) and lowest ($3.33 \times 10^7$ metric tons/year) amount of suspended sediment transport are measured in the months of July and March at Geonkhali and Kolaghat respectively. But in low tide condition, highest ($2.37 \times 10^8$ metric tons/year) and lowest ($2.26 \times 10^7$ metric tons/year) amount of suspended sediment transport are measured in the months of June and February at Geonkhali and Kolaghat respectively (Table 5.2). Average maximum suspended sediment transport is measured at Geonkhali in all the seasons except in post-monsoon high tide. During pre-monsoon high and low tide, average maximum sediment transports are measured as $9.8 \times 10^7$ and $8.3 \times 10^7$ metric tons/year respectively. During monsoonal high and low tide the average maximum transport are $2.45 \times 10^8$ and $2.3 \times 10^8$ metric tons/year, whereas in post-monsoon high and low tide, the average maximum transport are measured as $1.29 \times 10^8$ and $1.0 \times 10^8$ metric tons/year respectively (Table 5.2 and Fig. 5.1). The lowest sediment transport is measured as $2.5 \times 10^7$, $7.3 \times 10^7$ and $3.5 \times 10^7$ metric tons/year in pre-monsoon, monsoon and post-monsoon seasons respectively (Table 5.2). During monsoon period the amount of sediment penetrating inward during high tide is almost equal to the sediment that is drained back to ocean during low tide. Because of this the sedimentation rate is reduced in monsoon season (Table 1.2). But in non-monsoon season more sediment penetrates towards upstream during high tide than the amount of sediment discharging towards the sea during low tide (Table 5.2 and Fig. 5.1). Hence the region experiences a net input of sediment and accelerates the rate of sedimentation in this season (Table 1.2). Variation of suspended sediment transport towards downstream is mainly due to the

**Table 5.2** Amount of suspended sediment transport

| Month | Suspended sediment transport ('00000 metric tons/year) | | | | | | | | | | | |
|---|---|---|---|---|---|---|---|---|---|---|---|---|
| | Kolaghat | | Soyadighi | | Anantapur | | Pyratungi | | Dhanipur | | Geonkhali | |
| | High tide | Low tide | High tide | Low tide | High tide | Low tide | High tide | Low tide | High tide | Low tide | High tide | Low tide |
| February | 345 | 226 | 373 | 231 | 408 | 291 | 496 | 296 | 936 | 589 | 1020 | 865 |
| March | 333 | 252 | 361 | 292 | 373 | 298 | 521 | 278 | 831 | 533 | 995 | 803 |
| April | 385 | 236 | 402 | 251 | 362 | 278 | 474 | 361 | 817 | 547 | 947 | 807 |
| May | 423 | 286 | 426 | 310 | 419 | 284 | 547 | 386 | 908 | 611 | 963 | 845 |
| **Average (Pre-monsoon)** | **371** | **250** | **390** | **270** | **390** | **287** | **510** | **330** | **870** | **570** | **980** | **830** |
| June | 795 | 733 | 738 | 772 | 1070 | 995 | 1550 | 1150 | 2040 | 1720 | 2300 | 2370 |
| July | 715 | 747 | 766 | 733 | 1090 | 1070 | 1350 | 1500 | 2010 | 2230 | 2750 | 2350 |
| August | 716 | 745 | 774 | 715 | 1020 | 997 | 986 | 1250 | 1720 | 2130 | 2650 | 2230 |
| September | 774 | 735 | 722 | 690 | 1020 | 1050 | 1000 | 995 | 1580 | 1640 | 2100 | 2250 |
| **Average (Monsoon)** | **750** | **740** | **750** | **730** | **1050** | **1030** | **1190** | **1150** | **1950** | **1930** | **2450** | **2300** |
| October | 478 | 386 | 485 | 412 | 565 | 480 | 725 | 624 | 1650 | 870 | 1720 | 1540 |
| November | 462 | 372 | 483 | 386 | 465 | 408 | 605 | 536 | 1180 | 755 | 996 | 985 |
| December | 415 | 326 | 439 | 346 | 447 | 378 | 597 | 518 | 1190 | 752 | 998 | 984 |
| January | 405 | 323 | 433 | 338 | 445 | 374 | 593 | 522 | 1140 | 747 | 1020 | 979 |
| **Average (Post-monsoon)** | **440** | **350** | **460** | **370** | **480** | **410** | **630** | **550** | **1290** | **780** | **1170** | **1000** |

(*Source* Field measurement)

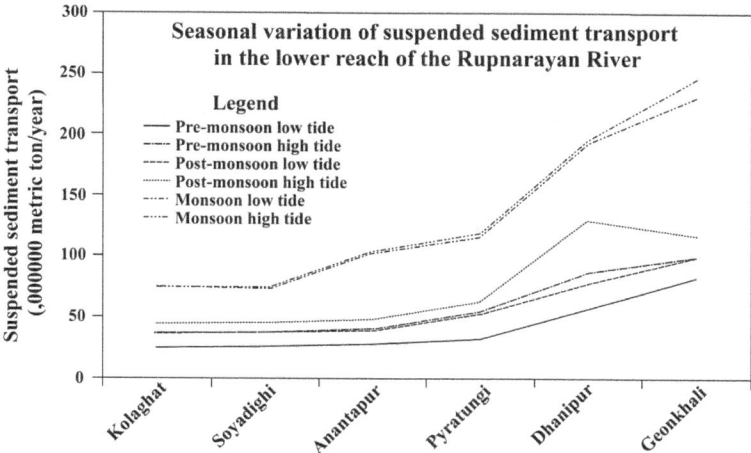

**Fig. 5.1**  Fluctuation of suspended sediment transportation

fluctuation of stream velocity and water discharge (Maity and Maiti 2012). Colby (1957) identified the mean stream velocity to be the most important variable influencing amount of sediment transport.

## 5.4.2   Dependence of Suspended Load Transport on Suspended Sediment Concentration

In the study area the relationship between amount of suspended sediment in water and suspended sediment transport is moderately positive ($r = 0.553$) (Fig. 5.2). Amount of suspended load not only depends on sediment concentration but also on other factors like, stream velocity and water discharge etc. If these factors remain same then the relation between suspended sediment concentration and transport will be absolutely positive. In the lower reach, the difference of suspended sediment concentration in water between non-monsoon and monsoon season and high and low tide water leads to the seasonal variation of suspended sediment transport and also its variation in both the tidal phases (Tables 5.1 and 5.2). In non-monsoon season the average concentration of suspended sediment in water is less (3.1–5.05 gm/l) than in monsoon season (4.97–6.5 gm/l), that's why the average suspended sediment transport is less in non-monsoon season ($2.5 \times 10^7$ to $1.17 \times 10^8$ metric tons/year) than monsoon season ($7.3 \times 10^7$ to $2.45 \times 10^8$ metric tons/year) (Tables 5.1 and 5.2). In dry season higher concentration of suspended sediment in high tide water increases the suspended sediment transport rate during high tide than in low tide condition (Tables 5.1 and 5.2). In pre-monsoon high tide, average volume of water discharge is equal (2750 m³/sec) at Anantapur and Pyratungi, but the average suspended sediment concentration is more at Pyratungi (4.1 gm/l) than at Anantapur (3.65 gm/l) (Table 5.1). Because of this the average suspended sediment transport is

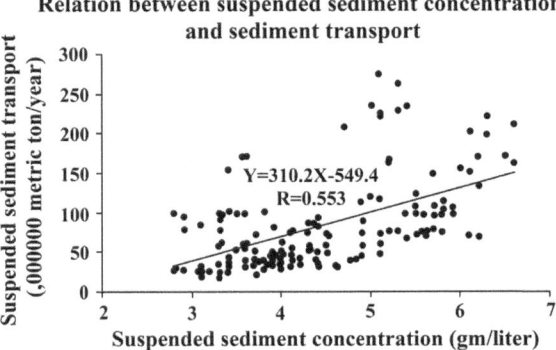

**Fig. 5.2**  Co-relation between suspended sediment concentration and transport

more at Pyratungi ($5.1 \times 10^7$ metric tons/year) than at Anantapur ($3.9 \times 10^7$ metric tons/year) (Table 5.2).

In natural stream, seasonal and spatial variation of water velocity and discharge alter the perfect positive relation between suspended sediment concentration and sediment transport. For instance, in all the seasons though the suspended sediment concentration is highest at Dhanipur but highest sediment transport is observed at Geonkhali mainly because of more water discharge at Geonkhali.

### 5.4.3  Dependence of Suspended Load Transport on Water Discharge

The relationship between *suspended sediment transport* and *water discharge* in natural stream is positive, though the degree of relationship varies spatially and temporally depending on situation. In the lower reach of the Rupnarayan River, the relationship between these two variables is strongly positive ($r = 0.908$) (Fig. 5.3), which reveals that the amount of suspended sediment transport is largely controlled by the volume of water discharge when other factors remain unchanged. In the month of May, low tide water discharge (940–3840 m$^3$/sec) and sediment transport ($2.86 \times 10^7$–$8.45 \times 10^7$ metric tons/year) are less in all the places, but in the month of June, sudden increase of water discharge (3300–9210 m$^3$/sec) leads to voluminous increase of suspended sediment transport ($7.33 \times 10^7$–$2.37 \times 10^8$ metric tons/year) (Tables 3.3 and 5.2). In all the months of non-monsoon (pre-monsoon and post-monsoon) season, the discharge of water is less compared to the discharge in the months of monsoon season (Tables 3.3) which causes less amount of suspended sediment transport in non-monsoon than in monsoon season (Table 5.2). For example, near Kolaghat, the average volume of water discharge and sediment transport during pre-monsoon low tide are 850 m$^3$/sec and $2.5 \times 10^7$ metric tons/year respectively (Tables 3.3 and 5.2). But in monsoon low tide due to increase of average water discharge (3455 m$^3$/sec) by monsoonal

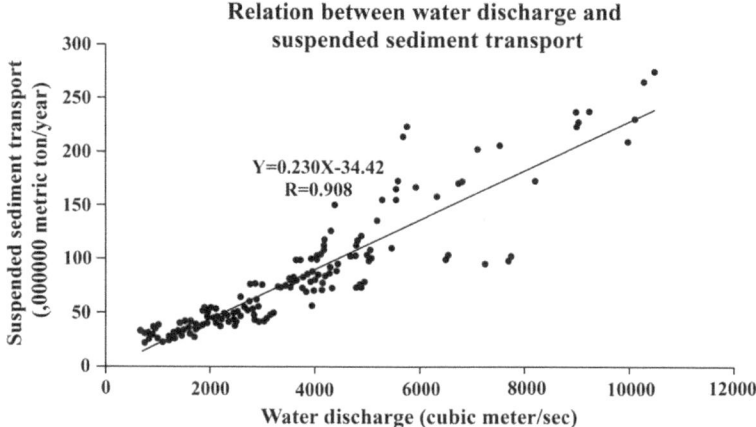

**Fig. 5.3** Co-relation between water discharge and suspended sediment transport

rainfall, the average suspended sediment transport has increased to $7.4 \times 10^7$ metric tons/year (Table 3.3 and 5.2). Again, during low tide of post-monsoon season the average water discharge at Kolaghat is 1020 m$^3$/sec and it has increased to 4160 m$^3$/sec at Geonkhali (Table 3.3), causing the increase of average suspended sediment transport from $3.5 \times 10^7$ metric tons/year at Kolaghat to $1.0 \times 10^8$ metric tons/year at Geonkhali (Table 5.2). Numerous studies have revealed that though the amount of suspended sediment transport depends on water discharge but actually the supply of sediments is closely related to surface run-off (Paola and Seal 1995). Generally, at a particular station the rate of increase of suspended load is more than the increase of water discharge. For instance, during pre-monsoon low tide the average discharge of water at Dhanipur is 2750 m$^3$/sec and it has increased to 5650 m$^3$/sec (almost two times) in monsoon low tide (Table 3.3). But the amount of suspended sediment transport has increased from $5.7 \times 10^7$ to $1.93 \times 10^8$ metric tons/year (almost four times) (Table 5.2), which indicates that the rate of increase of sediment transport is greater than the increase of water discharge. It has also been observed that water discharge and sediment transport do not always increase or decrease at the same time. During storm events the suspended sediment load reaches to peak before the water does. Lane and Borland (1951) stated that it is probably because of the fact that most of the loose debris lying on the surface are quickly carries away by the initial storm run-off.

## 5.5 Transportation of Bed Material Load

Bed load (generally >0.062 mm in diameter) is the clastic (particulate) material which moves through the channel completely supported by the river bed itself (Yeh et al. 1995). The bed materials, mostly the sand and gravel particles, are kept in motion (rolling and sliding) by means of the shear stress working at the channel boundary (Sundborg 1956). The amount of bed-load material in a river is always capacity limited i.e., it is a function of stream hydraulics rather than the supply of sediments (Swamee et al. 1991). Direct measurement of the amount of bed load transport in a tidal river is very difficult. The amount of bed load transport, at different stations in the lower reach of the Rupnarayan River has been calculated following different well accepted empirical equations of Meyer-Peter-Muller (1948), Einstein (1950), Yalin (1963), Bagnold (1966), Engelund-Hansen (1967), Yang (1973), Ackers-White (1973), Van Rijn (1984) and Samaga et al. (1986a). The results obtained from the equations of different researchers indicate that the bed load transport rate in river varies widely. Numerous studies have revealed that though the bed load transport varies due to the variation of water discharge but it is very difficult to establish a reliable and regular relationship between bed load transport and water discharge (Bogardi 1974) as the bed load transport depends on large number of variables which vary spatially and temporally.

## 5.5.1 *Fluctuation of Bed Load Transportation*

The amount of bed load transport as estimated by empirical equations of different researchers indicated that in the study area the *bed load transport rate* varies from reach to reach in different seasons. Usually, an increasing trend of bed load transport is observed from Kolaghat to Geonkhali (towards downstream) (Tables 5.3, 5.4, 5.5, 5.6, 5.7, 5.8 and Fig. 5.4). It is mainly because of the increase of water velocity and discharge from upstream to downstream (Maity and Maiti 2012). Few exceptions are found during monsoon period, when the bed load transport is low at Soyadighi than some other places (Tables 5.3, 5.4, 5.5, 5.6, 5.7, 5.8 and Fig. 5.4). The bed load transport rate varies from 0.1905 to 6.52985 kg/m/sec, 0.5008 to 14.74893 kg/m/sec and 0.2318 to 6.31764 kg/m/sec during pre-monsoon, monsoon and post-monsoon seasons respectively (Tables 5.3, 5.4, 5.5, 5.6, 5.7, 5.8 and Fig. 5.4). The transport of bed load is less in pre-monsoon and post-monsoon seasons compared to monsoon season. In pre-monsoon and post-monsoon seasons the bed load transport is always more during high tide than in low tide condition, but during monsoon period the rate of bed load transport is almost equal in both the tidal phases (Tables 5.3, 5.4, 5.5, 5.6, 5.7, 5.8 and Fig. 5.4). So, in dry season, the amount of inward penetration of sediment is more compared to the downward discharge of sediments into the ocean, resulting into the rapid rate of sedimentation in this season. But the condition becomes different in monsoon season as most of the sediments carried by high tide water are effectively discharged downstream by low tide water and the rate of sedimentation is reduced (Table 1.2).

Each and every bed load transport equation has been experimented in the field on streams and the results have normally been found to be less consistent with actual field measurements (Colby and Hembree 1955; Bogardi 1974; Emmett and Leopold 1977). Emmett and Leopold (1977) suggested that the difference between

**Table 5.3** Transportation of bed load at Kolaghat

| Sediment transport (kg/m/sec) (Using Empirical Equation) | Pre-monsoon | | Monsoon | | Post-monsoon | |
|---|---|---|---|---|---|---|
| | High tide | Low tide | High tide | Low tide | High tide | Low tide |
| Meyer-Peter-Muller (1948) | 0.2374 | 0.1905 | 0.5185 | 0.5008 | 0.2529 | 0.2318 |
| Einstein (1950) | 1.28305 | 1.11594 | 8.39261 | 8.019624 | 1.33841 | 1.02845 |
| Yalin (1963) | 1.5387 | 1.2745 | 3.8345 | 3.8095 | 1.5870 | 1.4309 |
| Bagnold (1966) | 2.9742 | 2.6519 | 7.75185 | 7.41962 | 3.09562 | 2.89045 |
| Engelund-Hansen (1967) | 1.7527 | 1.5839 | 3.7845 | 3.7794 | 1.6941 | 1.4827 |
| Yang (1973) | 0.85934 | 0.72196 | 6.73276 | 6.71094 | 0.93425 | 0.74319 |
| Ackers-White (1973) | 0.7843 | 0.6239 | 1.5094 | 1.5162 | 0.8452 | 0.6839 |
| Van Rijn (1984) | 0.97349 | 0.94521 | 10.44271 | 10.20943 | 1.12854 | 1.06523 |
| Samaga et al. (1986a) | 1.2870 | 1.2056 | 2.9582 | 2.9428 | 1.3059 | 1.2685 |

**Table 5.4** Transportation of bed load at Soyadighi

| Sediment transport (kg/m/sec) (Using Empirical Equation) | Pre-monsoon | | Monsoon | | Post-monsoon | |
|---|---|---|---|---|---|---|
| | High tide | Low tide | High tide | Low tide | High tide | Low tide |
| Meyer-Peter-Muller (1948) | 0.2365 | 0.2179 | 0.5635 | 0.5539 | 0.2418 | 0.2363 |
| Einstein (1950) | 1.19724 | 0.96283 | 4.90271 | 4.84091 | 1.66301 | 1.39810 |
| Yalin (1963) | 1.5628 | 1.4850 | 3.7238 | 3.5972 | 1.6630 | 1.4529 |
| Bagnold (1966) | 3.18752 | 2.90867 | 6.55318 | 6.09825 | 2.97490 | 2.64953 |
| Engelund-Hansen (1967) | 1.6371 | 1.4489 | 3.3864 | 3.3284 | 1.7150 | 1.6830 |
| Yang (1973) | 2.19029 | 1.93092 | 3.87192 | 3.79504 | 1.73950 | 1.55278 |
| Ackers-White (1973) | 0.7548 | 0.6832 | 1.3967 | 1.4184 | 0.9137 | 0.8293 |
| Van Rijn (1984) | 1.94896 | 1.69030 | 4.89231 | 4.88198 | 2.07417 | 1.89634 |
| Samaga et al. (1986a) | 1.0528 | 0.9278 | 2.9946 | 2.9106 | 1.1849 | 1.0528 |

**Table 5.5** Transportation of bed load at Anantapur

| Sediment transport (kg/m/sec) (Using Empirical Equation) | Pre-monsoon | | Monsoon | | Post-monsoon | |
|---|---|---|---|---|---|---|
| | High tide | Low tide | High tide | Low tide | High tide | Low tide |
| Meyer-Peter-Muller (1948) | 0.2996 | 0.2593 | 0.6532 | 0.6519 | 0.3034 | 0.2568 |
| Einstein (1950) | 1.62962 | 1.44861 | 7.57469 | 7.77921 | 1.93874 | 1.57893 |
| Yalin (1963) | 1.9850 | 1.8237 | 4.3175 | 4.2847 | 2.4129 | 2.2284 |
| Bagnold (1966) | 4.13842 | 4.01209 | 7.42854 | 7.52198 | 4.39851 | 4.18749 |
| Engelund-Hansen (1967) | 2.6398 | 2.4417 | 4.2764 | 4.2289 | 2.5628 | 2.3179 |
| Yang (1973) | 1.45281 | 1.20948 | 5.82910 | 5.810926 | 1.93829 | 1.63102 |
| Ackers-White (1973) | 1.3861 | 1.0638 | 2.6683 | 2.6492 | 1.4178 | 1.1109 |
| Van Rijn (1984) | 2.85482 | 2.55387 | 10.84563 | 10.80428 | 3.18749 | 3.01876 |
| Samaga et al. (1986a) | 1.7347 | 1.5189 | 6.7345 | 6.4297 | 2.1845 | 1.8274 |

**Table 5.6** Transportation of bed load at Pyratungi

| Sediment transport (kg/m/sec) (Using Empirical Equation) | Pre-monsoon | | Monsoon | | Post-monsoon | |
|---|---|---|---|---|---|---|
| | High tide | Low tide | High tide | Low tide | High tide | Low tide |
| Meyer-Peter-Muller (1948) | 0.3317 | 0.2952 | 0.6189 | 0.6098 | 0.3783 | 0.3551 |
| Einstein (1950) | 3.31854 | 3.09564 | 8.55218 | 8.56926 | 3.11964 | 2.79174 |
| Yalin (1963) | 2.7437 | 2.4184 | 5.2843 | 5.0937 | 1.9952 | 1.7318 |
| Bagnold (1966) | 5.10296 | 4.89538 | 8.55239 | 8.50942 | 4.78965 | 4.49631 |
| Engelund-Hansen (1967) | 2.2284 | 2.0157 | 4.5129 | 4.4942 | 2.1783 | 1.9420 |
| Yang (1973) | 2.88493 | 2.65940 | 5.66301 | 5.57352 | 3.15302 | 3.01923 |
| Ackers-White (1973) | 1.5594 | 1.2854 | 2.9531 | 2.8539 | 1.5673 | 1.3926 |
| Van Rijn (1984) | 3.21768 | 2.99736 | 9.52564 | 9.551875 | 2.95428 | 2.70934 |
| Samaga et al. (1986a) | 3.1843 | 2.7319 | 6.5290 | 6.3417 | 3.6215 | 3.2635 |

**Table 5.7** Transportation of bed load at Dhanipur

| Sediment transport (kg/m/sec) (Using Empirical Equation) | Pre-monsoon | | Monsoon | | Post-monsoon | |
|---|---|---|---|---|---|---|
| | High tide | Low tide | High tide | Low tide | High tide | Low tide |
| Meyer-Peter-Muller (1948) | 0.3173 | 0.2978 | 0.6381 | 0.6229 | 0.3851 | 0.3719 |
| Einstein (1950) | 2.74987 | 2.49841 | 7.553876 | 7.487590 | 3.12765 | 3.08387 |
| Yalin (1963) | 3.6520 | 3.1846 | 6.3197 | 6.0724 | 3.1893 | 2.9317 |
| Bagnold (1966) | 5.74291 | 5.48932 | 9.97473 | 9.76953 | 6.31764 | 6.09837 |
| Engelund-Hansen (1967) | 3.5286 | 3.3178 | 5.4271 | 5.4017 | 3.2176 | 3.0128 |
| Yang (1973) | 3.11074 | 3.10483 | 6.83920 | 6.82983 | 3.59483 | 3.38920 |
| Ackers-White (1973) | 2.0641 | 1.8041 | 3.8849 | 3.8528 | 1.9963 | 1.7157 |
| Van Rijn (1984) | 4.21785 | 4.01874 | 11.63782 | 11.61942 | 3.92671 | 3.63973 |
| Samaga et al. (1986a) | 3.1740 | 2.8451 | 6.7429 | 6.5831 | 3.8523 | 3.5318 |

**Table 5.8** Transportation of bed load at Geonkhali

| Sediment transport (kg/m/sec) (Using Empirical Equation) | Pre-monsoon | | Monsoon | | Post-monsoon | |
|---|---|---|---|---|---|---|
| | High tide | Low tide | High tide | Low tide | High tide | Low tide |
| Meyer-Peter-Muller (1948) | 0.4527 | 0.4286 | 0.8126 | 0.8011 | 0.3981 | 0.3812 |
| Einstein (1950) | 2.94710 | 2.59381 | 8.763023 | 8.690305 | 2.693829 | 2.38503 |
| Yalin (1963) | 3.6418 | 3.2189 | 6.6673 | 6.5092 | 3.8516 | 3.5519 |
| Bagnold (1966) | 6.52985 | 6.23973 | 11.62984 | 11.05983 | 6.29631 | 6.04297 |
| Engelund-Hansen (1967) | 3.9527 | 3.7593 | 5.2628 | 5.1942 | 3.7731 | 3.6187 |
| Yang (1973) | 2.55839 | 2.19820 | 7.94610 | 7.881014 | 3.18290 | 2.90819 |
| Ackers-White (1973) | 2.4739 | 2.2851 | 4.5593 | 4.4852 | 2.3834 | 2.1958 |
| Van Rijn (1984) | 3.18749 | 3.05318 | 14.74893 | 14.69838 | 4.52874 | 4.18795 |
| Samaga et al. (1986a) | 3.0749 | 2.8319 | 8.0348 | 7.9340 | 3.3965 | 3.0238 |

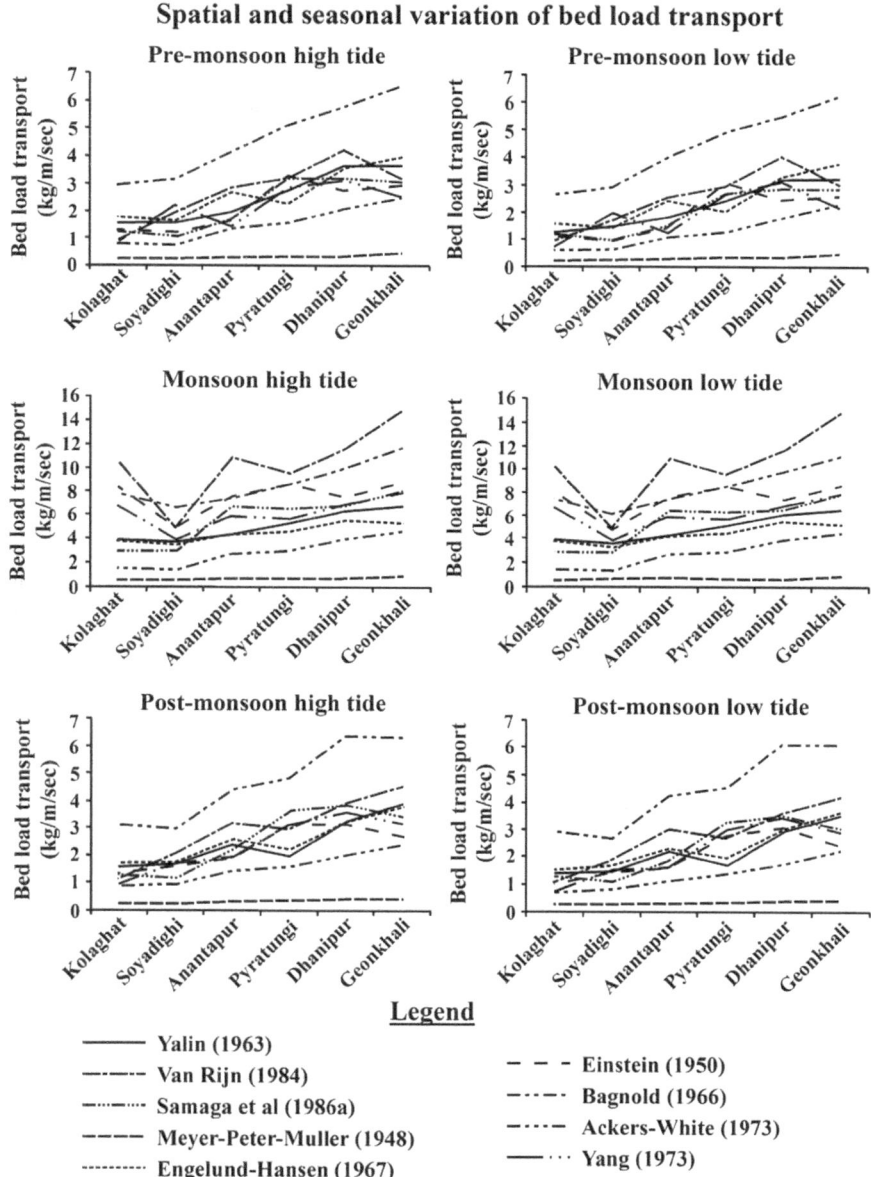

Fig. 5.4 Variation of bed load transportation

actual field measurements and estimated amount of bed load transport is mainly because of the variation of stream velocity, water discharge, river depth, width, bed slope, channel roughness and armouring etc. A river may transport a particular amount of sediment for different values of water velocity, bed slope, depth or water discharge. Similarly, a river may transport different amount of bed load for a fixed set of flow characteristics (Dunkerley 1990). The amount of bed load transport will change with the change of water velocity, a significant hydraulic variable and again the water velocity can be controlled by different other variables like, river depth and surface roughness (Parker et al. 1982). The presence of armour layer on river bed restricts the erosion and the supply of sediments so that the actual transport rate, particularly during marginal conditions, is lesser than that estimated by different *bedload formulae* (Dunkerley 1990). In case of bimodal sediment beds (coarse and fine strips) the transport rate is dissimilar; the coarser strips transport a smaller amount whereas the finer ones transport a larger amount of sediments than a river bed without such strips (Paola and Seal 1995). After evaluation of the performance of ten bed-load formulae in the field, Gomez and Church (1989) concluded that none of the formula provides satisfactory prediction of the bed-load transport rate. Though most of the researchers prefer Bagnold type *Stream power equations* but the results of these formulae are very insignificant and poor (Carson 1987). Garde and Ranga Raju (2000) tested these bed load formulae on different rivers and recommended that the formulae put forwarded by Engelund-Hansen (1967), Yang (1973) and Ackers-White (1973) are more applicable in the field because these formulae provide better results than others.

# References

Ackers P, White WR (1973) Sediment transport: new approach and analysis. J Hydraul Divn 99 (11):2041–2060

Allan RJ (1986) The role of particulate matter in the fate of contaminants in aquatic ecosystems. Inland Water Directorate Scientific Series, vol 142. Nat Water Resou Inst, Burlington, Canada

Andrews ED (1983) Entrainment of gravel from naturally sorted riverbed material. Geol Soc Amer Bull 94:1225–1231

Ashida K, Michiue M (1971) An investigation of river bed degradation downstream of a dam, Proc. 14th Congress of the IAHR

Bagnold RA (1966) An approach to the Sediment Transport Problem from general Physics. US Geol Survey Prof Paper 422:37

Bogardi J (1974) Sediment transport in alluvial streams. Akademiai Kiado, Budapest

Bridge JS, Bennett SJ (1992) A model for the entrainment and transport of sediment grains of mixed sizes, shapes and densities. Water Resour Res 28(2):337–363

Carson MA (1987) Measures of flow intensity as predictors of bed load. J Hydrol 113:1402–1421

Cobly BR (1957) Relationship of unmeasured sediment discharge to stream flow. US Geol Survey Open-File Report

Colby BR (1964) Scour and fill in sand bed streams. US Geol Survey Prof Paper, 462-D

Colby BR, Hembree CH (1955) Computations of total sediment discharge, Niobrara River near Cody, Nebraska. U S Geol Surv Water Suppl Pap, p 1357

Dunkerley DL (1990) The development of Armour in the Tambo River, Victoria, Australia. Earth Surf Process Landf 15:405–415

Dyer K (1973) Estuaries: a physical introduction. Wiley, London

Egiazaroff IV (1965) Calculation of non-uniform sediment concentration. J Hydraul Div Proc ASCE 106(HY4):1325–1343

Einstein HA (1942) Formulas for transportation of bed load. Trans ASCE 2140:561–597

Einstein HA (1950) The bed—load function for sediment transportation in open channel flows. USDA Tech Bull 1026(9):1978

Emmett WW, Leopold LB (1977) A comparison of observed sediment-transport rates computed using existing formulas. In: Doehring D (ed) Geomorphology in arid regions, (publications in geomorphology). SUNY-Binghamton, New York, pp 187–188

Engelund F, Hansen E (1967) A monograph for sediment transport and alluvial streams. Teknisk Forlag, Copenhagen

Fang D, Yu GL (1998) Bedload transport in cobble-bed rivers. Proc of International Water Resources Engineering Conference, Memphis, USA

Fournier F (1960) Climat et Erosion; la Relation entre l'Erosion du Sol l' Eau et les Precipitations Atmospheriques. Press University, France, Paris

Garde RJ, Ranga Raju KG (2000) Mechanics of sediment transport. New age International Publishers, New Delhi

Gilbert GK (1914) The transportation of debris by running water. US Geol Survey Prof Paper 86:259

Gomez B, Church M (1989) An assessment of bed load sediment transport formulae for gravel bed rivers. Water Resour Res 25:1161–1186

Hayashi TS, Ichibashi T (1980) Study on bed load transport of sediment mixture. Proc. 24th Japanese Conference on Hydraulics

Hickin EJ (1995) River geomorphology. Wiley, Chichester

Hsu EM, Holly FM (1992) Conceptual bed-load transport model and verification for sediment mixtures. J Hydraul Eng, ASCE 118(8):1135–1152

Karim F (1998) Bed material discharge prediction for non-uniform bed sediments. J Hydraul Eng, ASCE 124(6):597–604

Lane EW, Borland WM (1951) Estimating bedload. Amer Geophys Union Trans 32(1):121–123

Leopold LB, Wolman MG (1953) Fluvial processes in geomorphology. Freeman & Co, San Francisco, p 522

Maity SK, Maiti RK (2012) Impact of sedimentation on development and shifting of shoal area, pools and riffles and thalweg position at lower reach of Rupnarayan River-A case study. Indian J Power River Val Dev 24:46–54

Meybeck M (1993) C, N, P and S in rivers: from sources to global inputs. interactions of C, N. P and S biogeochemical cycles and global change. Springer, Berlin, pp 163–194

Meybeck M (1999) Global transfer of carbon by rivers, Global Change News Letter, 37. Stockholm, IGBP Secretariat, p 1820

Meyer-Peter E, Muller R (1948) Formulas of Bed Load Transport, Proc Third Meeting of IAHR, Stockholm, pp 39–64

Milliman JD, Meade RH (1983) Worldwide delivery of river sediments to the oceans. J Geol 91:1–21

Paola C, Seal R (1995) Grain-size patchiness as a cause of selective deposition and downstream fining. Water Resour Res 31:1395–1407

Parker G, Kilingeman PC, Mclean DG (1982) Bed load and size distribution in paved gravel-bed streams. J Hydraul Divn, ASCE 108(4):544–571

Proffit GT, Sutherland AJ (1983) Transport of non-uniform sediment. J Hydraul Res 21(1):33–43

Samaga BR, Ranga raju KG, Garde RJ (1986a) Bed load transport of sediment mixtures. J Hydraul Eng 112(11):1003–1018

Samaga BR, Ranga raju KG, Garde RJ (1986b) Suspended load transport rate of sediment mixture. J Hydraul Eng 11:1019–1038

Sheppard JR (1965) Methods and their suitability for determining total sediment quantities. In: Proceeding of the Federal Inter-Agency Sedimentation Conference, 1963, Misc Pub no. 970

Sundborg A (1956) The river Klaralven, a study on fluvial processes. Geogr Annals 38:127–316

Swamee PK, Ojha CSP (1991) Bed load and suspended load transport of non-uniform sediments. J Hydraul Eng, ASCE 117(6):774–787

Van-Rijn LC (1984) Sediment transport, Part II: Suspended load transport. J Hydraul Eng, ASCE 110(10):1613–1641

Walling DE (1989) Physical and chemical properties of sediment: the quality dimension. Int J Sed Res 4:27–29

Wilcock PR, McArdell BW (1993) Surface—based fractional transport rates: mobilization thresholds and partial transport of sand-gravel sediment. Water Resour Res 29:1297–1312

Yalin MS (1963) An expression for bed-load transportation. Proc ASCE 89:221–250

Yang CT (1973) Sediment transport: theory and practice. McGraw Hill International Edition, New York

Yeh KC, Li SJ, Chen WL (1995) Modeling non—uniform sediment fluvial process by characteristics methods. J Hydraul Eng, ASCE 121(2):159–170

# Chapter 6
# Analysis of Bed Load Sediment Texture

**Abstract** Textural analysis of surface sediments is very useful tool to understand the complex interaction between terrestrial and marine environment. Total 180 sediment samples (60 samples in each season) have been collected from the lower reach of the Rupnarayan River and sieving technique is used to calculate different size parameters. Approximately, 63.80% of the sediments are very fine sand, 14.76% are fine sand and 21.44% are coarse silt type. Sediments are coarser in monsoon than in pre-monsoon and post-monsoon seasons due to increase of water volume, stream energy and removal of fine sediments in monsoon. In dry season, >60% sediments are moderately to well sorted but in monsoon season 63.85% sediments are poorly to very poorly sorted. Around 55% of the sediments are of fine and very fine skewed type, 33% of samples are near symmetrical and remaining are of coarse skewed type. The coarser sediments are negatively skewed and finer sediments are positively skewed. In monsoon, >60% of the sediments is platykurtic or leptokurtic in nature which indicates the high energy environment in this season. Proportion of sand, silt and clay in sediments ranges between 38–91%, 4–61% and 1–41% respectively. Nearly, 81.33% of the sediments are silty sand, 7.33% are muddy sand and 6% samples are of sandy silt category.

**Keywords** Grain size · Sorting · Skewness · Kurtosis · Sediment type

## 6.1 Introduction

The *sedimentological analysis* of undisturbed surface sediments is very useful tool (Rashi et al. 2011) to understand and explain the mechanism of complex dynamic systems in the zone of transition between terrestrial and marine environment. The characteristics of the bed sediment are very significant, for it is not only the size of the sediment, but also for the association of different grain sizes (Dietrich et al. 1989; Buffington and Montgomery 1997; Church 2006). Sedimentologists and geologists thought that cumulative curves of sediment size distribution represent the distinct subpopulations of sediment with a lognormal distribution, which reflect

© The Author(s) 2018
S. Kumar Maity and R. Maiti, *Sedimentation in the Rupnarayan River*,
SpringerBriefs in Earth Sciences, https://doi.org/10.1007/978-3-319-62304-7_6

different modes of sediment transportation (Visher 1969). Grain size distribution of estuarine sediments are greatly affected by different factors like, characteristic of the source area, climatic type, length and energy of sediment transport in the *depositional environments* (Bhatia and Cook 1986; Fralick and Kronberg 1997; Venkatramanan et al. 2010). Numerous studies have shown that sediment size decreases with distance of transportation, especially in the fluvial environment (Pettijohn 1975). Classification of sediment has been attempted by plotting the percentage of sand, silt and clay in a *triangular diagram* proposed by Folk (1974). The graphical and moment measures of Folk and Ward (1957) for statistical parameters (Mean, Median, Sorting, Skewness etc.) not only reflect the environment of deposition but also found helpful in identifying hydraulic conditions of transporting medium and depositional basin (McLaren and Bowels 1985; Ghosh and Chatterjee 1994). River sediments are sandy silt and coarse grained in nature, but the sediments of estuarine environment are clayey silt and fine grained (Muraleedharan Nair and Ramachandran 2002; Rashi et al. 2011). Inman (1952) and Friedman (1967) indicated that, generally fine sediments are *well sorted* and coarser sediments show the tendency to be *poorly sorted*. *Negative Skewness* (coarse skewness) is correlated with high energy and winnowing action (removal of fines) and *positive skewness* (fine skewness) with low energy levels (accumulation of fines) (Friedman 1961). Brambati (1969) mentioned that, the positive skewness is the indicative of unidirectional transportation of sediments (channel) or the sediments have been deposited in sheltered low energy environment. Friedman (1962) suggested that extreme high or low values of *kurtosis* imply that part of the sediment achieved its sorting elsewhere in a high energy environment (Baruch et al. 1997). The nature and characteristics of flow of the depositing medium is reflected by the variations of kurtosis values (Seralathan and Padmalal 1994).

## 6.2  Materials and Methodology of the Study

A total number of *180 sediment samples*, including 60 during pre-monsoon, 60 during monsoon and 60 during post-monsoon have been collected from different sections of the lower reach of the Rupnarayan River (Figs. 6.1 and 6.2a) based on grain size variation, color variation, geomorphic unit and area of shoaling and scouring using hand auger. Sediment samples were dried and carefully mixed (Fig. 6.2b, c). Dry sediment samples of 100 gm were kept in the topmost sieve of stacked sieves (Fig. 6.2d). The set of sieves was arranged in such a way that the coarsest sieve put at the top and continuously the finer sieves below. The sieve set was then kept on an electrical shaking machine. Then the sediment sample was sieved for 20 min using *sieves* at half phi intervals. The sediment remained on each sieve and pan was weighed and cumulative weight (in percentage) was calculated and then cumulative frequency curves were drawn (Maity 2015). All the testing and analysis related to grain size were performed in the laboratory of Central Research Facility (CRF) of IIT, Kharagpur, West Bengal, India. *Mean, Median, Sorting,*

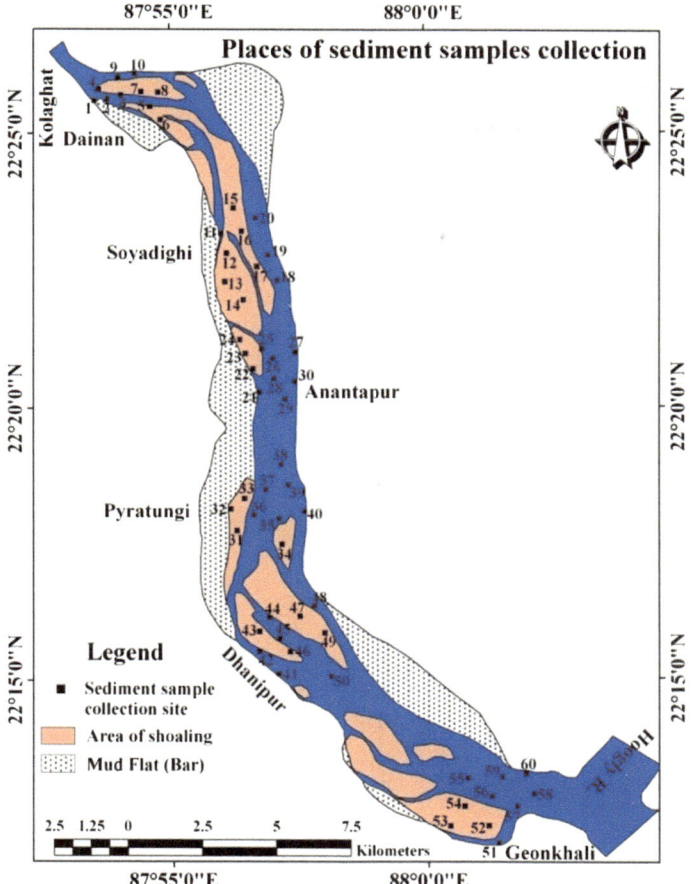

**Fig. 6.1** Sediment sample collection sites in the study area

*Skewness* and other grain size parameters are calculated (in phi units) using the conventional method of Folk and Ward (1957).

$$\text{Mean} = \frac{(\phi16 + \phi50 + \phi84)}{3} \tag{6.1}$$

$$\text{Standard Deviation} = \frac{(\phi84 - \phi16)}{4} + \frac{(\phi95 - \phi5)}{6.6} \tag{6.2}$$

$$\text{Skewness} = \frac{(\phi16 + \phi84 - 2\phi50)}{(2\phi84 - \phi16)} + \frac{(\phi5 + \phi95 - 2\phi50)}{(2\phi95 - \phi5)} \tag{6.3}$$

**Fig. 6.2** Collection of sediment samples (**a**), drying of sediments (**b**), preparation of sediments (**c**) and sieves used for grain size analysis (**d**)

$$\text{Kurtosis} = \frac{(\phi 95 - \phi 5)}{2.44 \, (\phi 75 - \phi 25)} \qquad (6.4)$$

Classification of sediment has been done by plotting the percentage of sand, silt and clay in a *triangular diagram* proposed by Folk (1980).

## 6.3  Sediment Texture at Different Reaches

Textural parameters of sediment like, Mean, Sorting, Skewness, and Kurtosis are intensively used to understand the depositional environments of sediments (Amaral and Prayor 1977). Numerous studies show the relation between sediment size parameters and transport processes and environment of sediment deposition (Rajgnapathi et al 2012; Folk and Ward 1957; Mason and Folk 1958; Friedman 1961 and 1967; Visher 1969; Valia and Cameron 1977; Wang et al. 1998; Asselman 1999). The lower reach of the Rupnarayan River is characterized by wide spatial and temporal variation of grain size parameters (like average grain size, nature of sorting, degree of skewness and value of Kurtosis), which is discussed under different reaches.

## *6.3.1   Mean Grain Size of Sediment*

*Mean size* of the sediments indicate the central tendency or the average size of the sediments and it indicates the available kinetic energy of the depositional agent (Sahu 1964). It is influenced by the source of supply, transporting medium and the energy conditions of the depositing environment (Visher 1969; Sly et al. 1982). The mean grain size at Kolaghat region varies between 2.50–3.25 $\phi$, 2.68–3.12 $\phi$ and 2.76–3.28 $\phi$ during pre-monsoon, monsoon and post-monsoon respectively (Fig. 6.3a–c). More than half (60%) of the studied sediment samples fall in fine sand category (mainly found in monsoon season) and remaining 40% are of very fine sand category.

At Soyadighi the mean grain size varies between 2.89–3.51 $\phi$, 2.51–3.192 $\phi$ and 2.88–3.62 $\phi$ during pre-monsoon, monsoon and post-monsoon respectively (Fig. 6.3a–c). More than half (60%) of the studied sediment samples fall in very fine sand category and remaining 40% of the samples are fine sand type.

The mean grain size of sediments at Anantapur reach is comparatively finer than upper sections of the study area. It ranges between 3.33–4.27 $\phi$, 3.25–4.01 $\phi$ and 3.19–3.99 $\phi$ during pre-monsoon, monsoon and post-monsoon respectively (Fig. 6.3a–c). Nearly, 90% of the studied sediment samples fall in very fine sand category and remaining sediments are coarse silt in nature.

The proportion of silt and clay in sediment has increased to a great extent at Pyratungi reach. The mean grain size varies between 3.503–4.55 $\phi$, 3.34–4.15 $\phi$ and 3.58–4.32 $\phi$ during pre-monsoon, monsoon and post-monsoon respectively (Fig. 6.3a–c). Nearly 63.34% of the studied sediment samples fall in very fine sand category and 36.66% of the samples are coarse silt type, which indicates that the sediments are finer than the sediments of upper reaches.

At Dhanipur, the sediments are finer than the sediments of all the places of the study area. Reduction of stream energy at Dhanipur is the main reason behind it. The mean grain size varies between 3.85–4.98$\phi$, 3.421–4.58 $\phi$ and 3.58–4.873 $\phi$ during pre-monsoon, monsoon and post-monsoon respectively (Fig. 6.3a–c). More than half (60%) of the studied sediment samples fall in coarse silt category and 40% of the samples are very fine sand type. More than 70% of the sediment samples in monsoon season are of very fine sand type.

The mean grain size, at Geonkhali varies between 3.28–4.98 $\phi$, 3.18–4.02 $\phi$ and 3.33–4.38 $\phi$ during pre-monsoon, monsoon and post-monsoon respectively (Fig. 6.3a–c). More than half (60%) of the studied sediment samples fall in very fine sand category and remaining 40% of the samples are coarse silt type.

The area under study is characterized by spatial and seasonal variation of distribution of sediment grain size. Nearly, 63.80% of the sediments are *very fine sand*, 14.76% are *fine sand* and remaining 21.44% are coarse silt in nature. During monsoon period mean grain size is coarser than pre-monsoon and post-monsoon seasons (Fig. 6.3a–c). It is mainly due to monsoonal rainfall, voluminous terrestrial discharge, accelerated soil erosion and huge supply of riverine sediment from upper catchment area during this season (Maity 2015). Voluminous riverine discharge

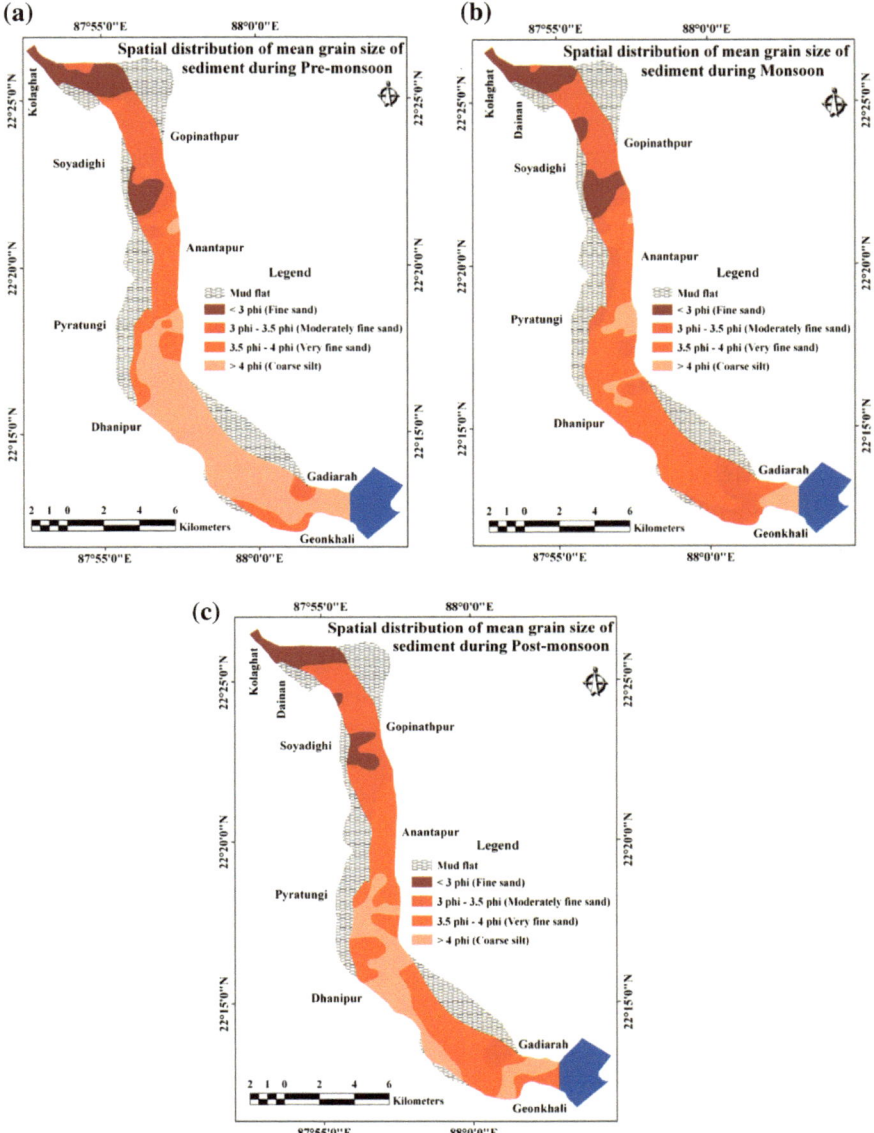

**Fig. 6.3  a** Spatial distribution of mean grain size during pre-monsoon. **b** Spatial distribution of mean grain size during monsoon. **c** Spatial distribution of mean grain size during post-monsoon. (*Source* Field survey and laboratory experiment)

enhances the stream energy and the fine sediments are removed easily, making the river bed sediments coarser. But during non-monsoonal season the paucity of riverine discharge reduces the stream energy during low tide causing the deposition of finer sediments. The spatial distribution of sediment grain size in the study area is haphazard and chaotic in nature. This is because of the interaction of fluvial and marine processes and the mixing of sediments during high and low tide have made the grain size distributional pattern more chaotic and haphazard. The sediments drained from upstream are caught up in the estuary and again redistributed upstream by stronger flood tide (Maity 2015). Though, no significant trend of spatial distribution of sediment grain size towards downstream or upstream is found but the sediments become slightly finer towards downstream. Generally, coarse silt and very fine sands were deposited at a low and moderately low energy conditions and the fine sand were deposited at a moderate energy conditions (Sahu 1964). Gradual and slight decrease of mean grain size towards downstream exhibits the gradual increase in energetic condition towards downstream. Again, the increase of marine influence towards downstream accelerates the mixing of fresh water with saline water and the suspended finer sediments settle down very slowly, while particles subject to electro-chemical reaction will flocculate upon contact with the salt water (Rhodes 1950), making the sediments slightly finer towards downstream.

### 6.3.2  Sorting of Sediment

The fluctuations in the kinetic energy and velocity conditions of the depositing agent can be easily understood by the calculation of *Standard deviation* of sediment grain size (Sahu 1964; Rashi et al. 2011). Standard deviation also measures the degree of uniformity of grain size within a sediment sample. It is one of the most useful parameter mediums in separating grains. Near Kolaghat region, the value of standard deviation varies from 0.63 to 1.67 $\phi$ in pre-monsoon, while in monsoon it varies from 0.82 to 1.68 $\phi$. During post-monsoon it ranges between 0.48 $\phi$ and 1.56 $\phi$ (Fig. 6.4a–c). In pre-monsoon and post-monsoon season most of the sediments (55%) are moderately to moderately well sorted, while in monsoon period nearly 70% samples are poorly sorted (Fig. 6.4a–c).

During pre-monsoon the value of standard deviation at Soyadighi varies from 0.45 to 1.52 $\phi$, while in monsoon it varies from 0.72 to 1.83 $\phi$. During post-monsoon it ranges between 0.59 $\phi$ and 1.99 $\phi$ (Fig. 6.4a–c). In pre-monsoon and post-monsoon season most of the sediments (>70%) are moderately to moderately well sorted, while in monsoon period more than 50% samples are poorly sorted.

Near Anantapur, the value of standard deviation varies from 0.72 to 2.02 $\phi$ in pre-monsoon, while in monsoon it varies from 0.62 $\phi$ to 1.85 $\phi$. During post-monsoon it ranges between 0.77 $\phi$ and 1.53 $\phi$ (Fig. 6.4a–c). More than 63% of the sediments are poorly sorted and remaining samples are moderately to moderately well sorted. In pre-monsoon season most of the sediments (66%) are moderately to moderately well sorted while in monsoon and post-monsoon seasons more than 75% samples are poorly sorted (Fig. 6.4a–c).

Near Pyratungi region, during pre-monsoon the value of standard deviation varies from 0.73 to 1.78 $\phi$, while in monsoon it varies from 0.67 to 2.15 $\phi$. During

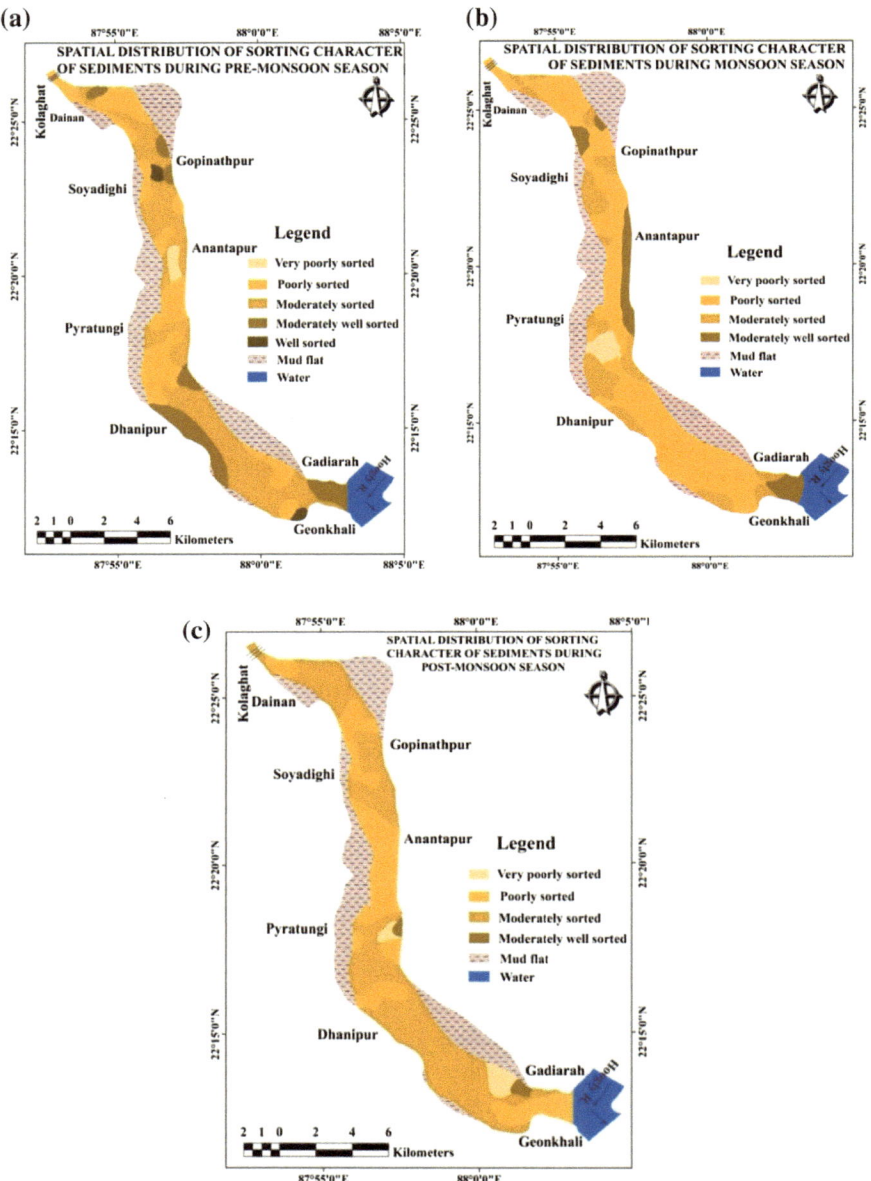

**Fig. 6.4**  **a** Spatial variation of sorting in pre-monsoon. **b** Spatial variation of sorting in monsoon. **c** Spatial variation of sorting in post-monsoon. (*Source* Field survey and laboratory experiment)

post-monsoon it ranges between 0.66 $\phi$ and 2.08 $\phi$. Nearly, 46.66% of the studied sediments are moderately to moderately well sorted and remaining 53.34% of the sediments are poorly to very poorly sorted. In monsoon season more than 70% sediments are poorly to very poorly sorted (Fig. 6.4a–c).

At Dhanipur, the value of standard deviation varies from 0.55 to 1.32 $\phi$ in pre-monsoon season, while in monsoon season it varies from 0.74 to 1.29 $\phi$. During post-monsoon it ranges between 0.74 $\phi$ and 1.37 $\phi$. In pre-monsoon and post-monsoon season most of the sediments (70%) are moderately to moderately well sorted, while in monsoon period more than 50% samples are poorly sorted (Fig. 6.4a–c).

During pre-monsoon the value of standard deviation at Geonkhali varies from 0.47 to 1.74 $\phi$, while in monsoon it varies from 0.62 to 1.77 $\phi$. During post-monsoon it ranges between 0.62 $\phi$ and 2.07 $\phi$ (Fig. 6.4a–c). Nearly, 53.33% of the sediment samples are moderately to moderately well sorted and remaining are poorly to very poorly sorted. More than 60% of the sediments in monsoon season are poorly or very poorly sorted.

In fluvial system the sediments are well sorted from upstream to downstream, i.e., sediments become finer with distance from source region, due to continuous reduction of stream velocity and energy. But in estuarine river the nature of sorting of sediments is quite different from the stream without estuarine influence. As the lower reach of the Rupnarayan River is affected by estuarine characteristics, the fluctuation of stream energy during high and low tide in different seasons has made the nature of sorting of sediment more complicated and unpredictable. During monsoon season, nearly 63.85% of the sediments are poorly to very poorly sorted and remaining are moderately to well sorted (Fig. 6.4b). But, during pre-monsoon and post-monsoon season more than 60% sediments are moderately to well sorted (Fig. 6.4a, c). Generally, the places where the sediments are coarse grained are characterized by poor sorting of sediments. In monsoon season, the coarseness of sediments than non-monsoon season is the main reason of the poor sorting of sediments. Inman (1952) and Friedman (1967) observed that coarser sediments tend to show deterioration in sorting whereas fine sediments are well sorted. Voluminous discharge of riverine water in monsoon season makes the low tide and high tide condition equally energetic, i.e., the sediments that are penetrating inward during high tide, again discharged towards downstream, making the sediments coarser and poorly sorted. But in non-monsoonal season, due to scarcity of terrestrial discharge the low tide is weaker causing the deposition of fine sediments and well sorting of sediments. Near Dhanipur reach most of the sediments are moderately to moderately well sorted due to fineness of the sediments (Fig. 6.4a–c). The mixing of sediments due to interaction of fluvial and marine processes is the main reason of the hapazard and unpredictable sorting of the sediments (Maity 2015).

### 6.3.3 Skewness

*Skewness* is an important parameter to measure the *asymmetry* of the grain size distribution in a sediment sample and it is considered as a sensitive indicator of sub-population mixing. According to the sedimentologists, skewness is an extremely sensitive measure indicating the type of transport or deposition agent of the

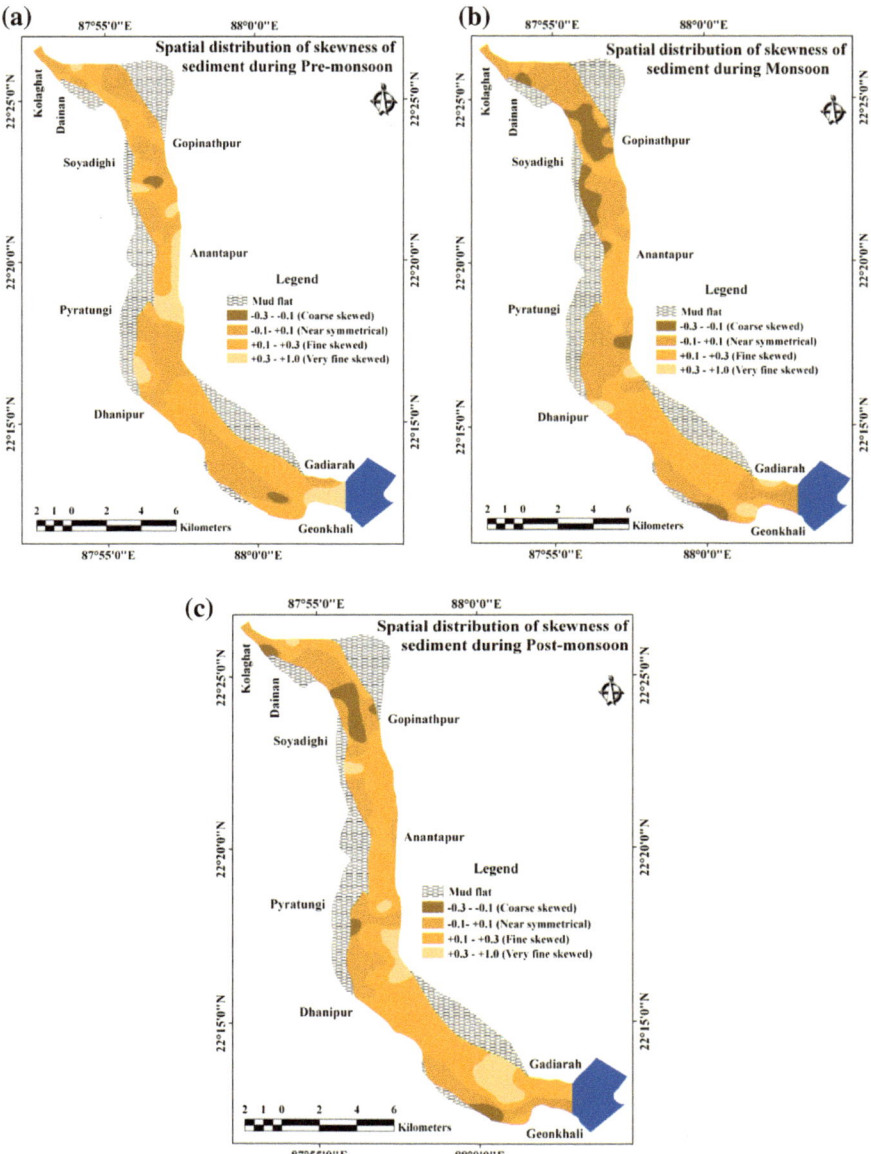

**Fig. 6.5 a** Spatial distribution of skewness during pre-monsoon. **b** Spatial distribution of skewness during monsoon. **c** Spatial distribution of skewness during post-monsoon. (*Source* Field survey and laboratory experiment)

sediment. Sign of skewness reveals the availability of energy in the environment of sediment deposition (Duane 1964). Near Kolaghat region, 7% of the sediment samples are very fine skewed type, 53% of the samples are of fine skewed type, 27% of samples are near symmetrical and remaining 13% samples are of coarse skewed type (Fig. 6.5a–c). Nearly 80% of the samples, collected during pre-monsoon and post-monsoon seasons are characterized by positive skewness but most of the samples (60%) of monsoon season are of negatively skewed.

At Soyadighi, nearly 54% of the sediment samples are of fine skewed or very fine skewed type, while 26% are of near symmetrical in nature and remaining 20% of samples are of coarse skewed (Fig. 6.5a–c). Most of the sediment samples (85%) collected during pre-monsoon and post-monsoon seasons are characterized by positive skewness but most of the samples (60%) of monsoon season are of negatively skewed.

At Anantapur, more than 60% of the sediments are near symmetrical type, while 36.66% of sediments are fine to very fine skewed and only 3.37% (one sample in monsoon season) samples are of coarse skewed type (Fig. 6.5a–c). Most of the samples (93.37%), collected during pre-monsoon and post-monsoon seasons are characterized by positive skewness but 40% of the sediment samples in monsoon season are of negatively skewed.

Nearly 40% of samples at Pyratungi reach are fine skewed to very fine skewed type, while 53.33% of the sediment samples are of nearly symmetrical type and remaining 6.66% (one sample in monsoon and post-monsoon in each) of the sediments are coarse skewed in nature (Fig. 6.5a–c). Most of the sediments (80%) being finer are characterized by positive skewness.

Near Dhanipur region, 56.66% of the sediment samples are of fine skewed to very fine skewed type and remaining 43.34% of samples are near symmetrical in nature. Most of the samples (86.66%) are characterized by positive skewness and only 13.33% samples are negatively skewed (Fig. 6.5a–c).

At Geonkhali, nearly 30% of samples are very fine skewed type, 40% of the sediment samples are of fine skewed type, while 20% samples are near symmetrical and remaining 10% samples are coarse skewed in nature. More than 80% of the sediments are positively skewed and only 20% are negatively skewed (Fig. 6.5a–c).

In the entire lower reach, around 55% of the sediment samples are of *fine and very fine skewed* type, while 33% of samples are *near symmetrical* and remaining are of coarse skewed type. Mainly the coarser sediments are negatively skewed (coarse skewed) and finer sediments are positively skewed (fine skewed). During non-monsoon (pre-monsoon and post-monsoon) season most of the sediments (>59%) are positively skewed (Fig. 6.5a, c) but in monsoon season most of the sediments (>60%) are negatively skewed (Fig. 6.5b). A mixture of positive and negative skewness of sediments is observed in all the seasons. The interaction of fluvial and marine processes, the fluctuation of energy during high and low tide and the mixing of sediments lead to such type of mixture of positive and negative skewness in the area under study. According to Martins (2003), negatively skewed curves are indication of non-depositional places, whereas positively skewed curves indicate deposition and a mixture of positive and negative skewness indicates an area in a state of flux. In monsoon season huge volume of terrestrial discharge increases stream energy and the finer materials are easily removed (scouring is more than shoaling) towards downstream causing the negative skewness of the sediments (Maity 2015). But in non-monsoonal season the scarcity of terrestrial discharge reduces stream energy, sediment transporting capacity and accelerates the rate of sedimentation by accumulating the fine materials. Friedman (1961) indicated that negative skewness is correlated with high energy condition and positive skewness with low energy levels (Awasthi 1970). Sahu (1964), Martins (1965 and 1967), Cronan (1972), Friedman (1961) indicated that negative skewness has relationship with the intensity and duration of high energy depositional agent through a removal of fines (Awasthi 1970). Duane (1964) elaborated that sign of skewness is related to energy variation and indicated that winnowing action induced by fluid media is the main mechanism producing negative skewness, whereas sediments deposited in sheltered environment are dominantly positively skewed. Valia and Cameron (1977) also agree that positive skewness is resulted from the addition of finer materials.

### 6.3.4   Kurtosis

The contrast between the sorting of the particle size distribution observed in the central part and that of the tails is understood by the measure of the *Kurtosis*. Duane (1964) and Mason and Folk (1958) mentioned that it is painstaking as one of the important grain size parameters to distinguish different depositional environments. At Kolaghat region the value of Kurtosis varies from 0.44 to 1.12, 0.552 to 1.098 and 0.56 to 1.19 during pre-monsoon, monsoon and post-monsoon respectively (Fig. 6.6a–c). Nearly, 47% of the sediment samples are mesokurtic in nature while 43% samples are platykurtic to very platykurtic in nature and remaining 10% are of leptokurtic in category.

Near Soyadighi, the value of Kurtosis varies from 0.674 to 1.59, 0.43 to 1.53 and 0.763 to 1.54 during pre-monsoon, monsoon and post-monsoon respectively (Fig. 6.6a–c). Nearly, 36.66% of the sediment samples are mesokurtic, 26.66% samples are leptokurtic and remaining 36.68% samples are platykurtic and very platykurtic in nature.

Near Anantapur, the value of Kurtosis varies from 0.65 to 1.231, 0.648 to 1.29 and 0.62 to 1.15 during pre-monsoon, monsoon and post-monsoon respectively (Fig. 6.6a–c). Nearly 46.66% of the sediment samples are mesokurtic in nature, while 36.66% of the samples are platykurtic and very platykurtic in nature. Remaining 16.68% samples are leptokurtic in nature.

Pyratungi reach is characterized by the Kurtosis values of 0.629–1.548 in pre-monsoon, 0.83–1.572 in monsoon and 0.81–2.012 in post-monsoon season (Fig. 6.6a–c). 50% of the sediment samples are mesokurtic in nature while 30% samples are leptokurtic and very leptokurtic in character. Remaining 20% of the samples are platykurtic to very platykurtic type.

At Dhanipur reach, the value of Kurtosis varies from 0.48 to 1.93 in pre-monsoon, 0.92 to 1.82 in monsoon and 0.43 to 1.85 in post-monsoon

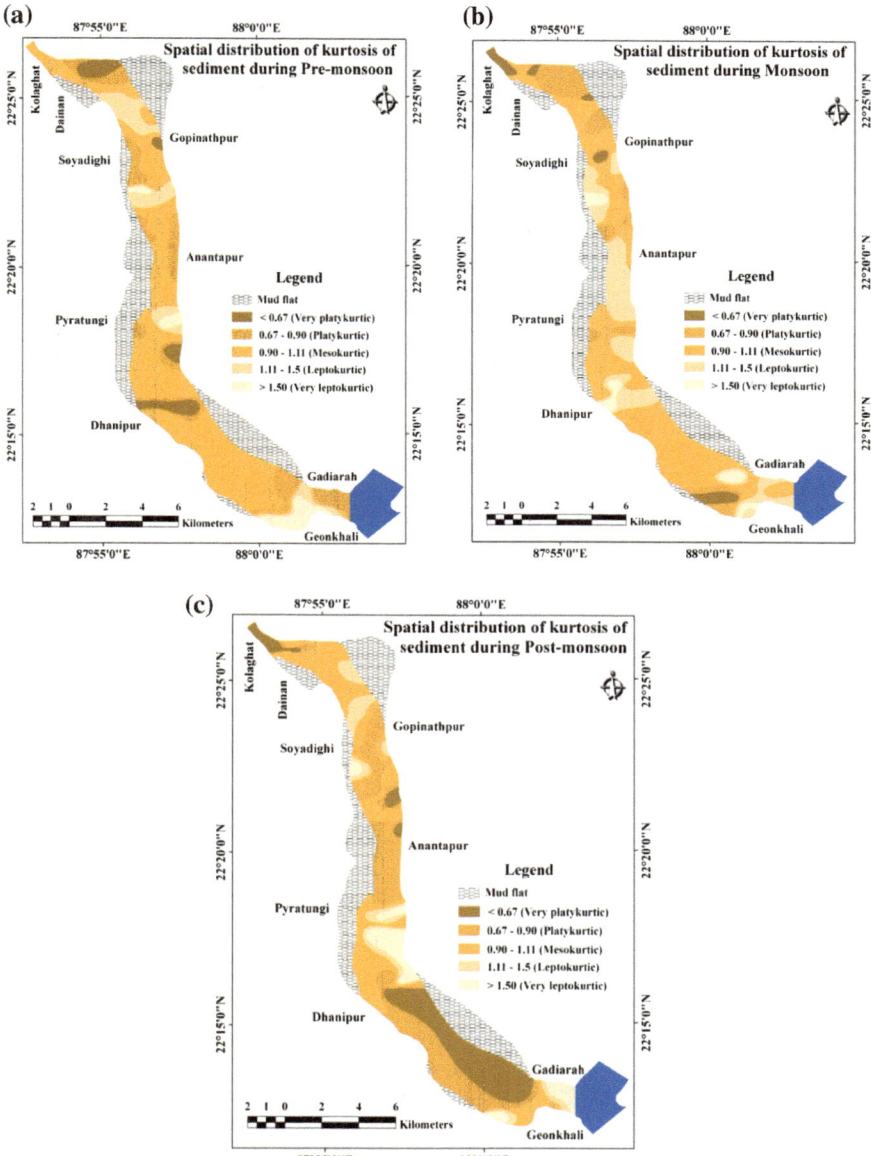

**Fig. 6.6** **a** Spatial distribution of kurtosis during pre-monsoon. **b** Spatial distribution of kurtosis during monsoon. **c** Spatial distribution of kurtosis during post-monsoon. (*Source* Field survey and laboratory experiment)

(Fig. 6.6a–c). 53.33% of the sediment samples are mesokurtic in nature while 36.66% of the samples are very leptokurtic to very platykurtic in nature. Remaining 10% samples are of leptokurtic type.

Near Geonkhali region, the value of Kurtosis varies from 0.64 to 1.73, 0.45 to 1.98 and 0.52 to 1.73 during pre-monsoon, monsoon and post-monsoon respectively (Fig. 6.6a–c). Nearly, 26.66% of the sediment samples are mesokurtic in nature while 40% samples are of leptokurtic and very leptokurtic in nature. Remaining 33.34% samples are platykurtic and very platykurtic in nature.

The variation in the kurtosis values is a reflection of the flow characteristics and the fluctuation of energy in the depositing environment (Seralathan and Padmalal 1994; Baruch et al. 1997). The entire lower reach is characterized by great variation of Kurtosis values in different seasons. More than 46% of the sediment samples are mesokurtic in nature, while very leptokurtic and very platykurtic samples are around 14% in each. Remaining 26% samples are leptokurtic and platykurtic in nature. During pre-monsoon season, more than 58% of the sediment samples are mesokurtic in nature (Fig. 6.6a), but in monsoon season more than 60% of the sediments are platykurtic or leptokurtic in nature (Fig. 6.6b), which indicates the high energy environment in monsoon season and comparatively low to medium energetic environment in pre-monsoon season. In post-monsoon season, 55% of the sediment samples are characterized by very high or low values of Kurtosis, indicating moderately high energetic condition in the depositing environment (Fig. 6.6c). According to Friedman (1962) very high and low values of kurtosis indicate that part of the sediments have sorted elsewhere in a high energy environment.

## 6.4   Proportion of Sand-Silt-Clay in Sediments

Considering all the sediment samples together the proportion of sand ranges between 38–91%, silt ranges between 4–61% and clay ranges between 1–41% (Table 6.1). During monsoon season the proportion of sand is more (52–91%) than non-monsoon season (38–91% in pre-monsoon and 45–86% in post-monsoon) (Table 6.1). The amount of silt varies between 8–61% and 4–54% in pre-monsoon and post-monsoon respectively, but it varies between 8–45% in monsoon season. Proportion of clay ranges between 1–41% in pre-monsoon, 1–19% in monsoon and 1–25% in post-monsoon season (Table 6.1) (Maity 2015). Particles are coarse in monsoon season than in non-monsoon season (pre-monsoon and post-monsoon). In monsoon season the increase of water volume leads to the increase of stream energy causing the easy removal of fine sediments and the sediments become coarse grained (Friedman 1961).

There is no conspicuous trend in the spatial distribution of sediment grain size either towards upstream or downstream at the lower reach. As the study area has estuarine characteristics, the interaction of fluvial and marine processes and the mixing of sediments during high and low tide have made the sediment grain size

**Table 6.1** Sand-silt and clay ratio in sediments

| Locations | Sand-silt-mud proportion (%) | | | | | | | | |
|---|---|---|---|---|---|---|---|---|---|
| | Pre-mnsoon | | | Monsoon | | | Post-monsoon | | |
| | Sand | Silt | Clay | Sand | Silt | Clay | Sand | Silt | Clay |
| Kolaghat | 68–91 | 8–30 | 1–15 | 76–91 | 8–17 | 1–12 | 71–86 | 9–20 | 5–20 |
| Soyadighi | 60–78 | 12–25 | 8–18 | 70–86 | 8–21 | 2–18 | 70–84 | 7–21 | 8–18 |
| Anantapur | 45–78 | 13–48 | 6–38 | 59–86 | 10–42 | 4–17 | 55–78 | 14–39 | 4–22 |
| Pyratungi | 54–79 | 8–32 | 12–26 | 73–87 | 9–20 | 4–15 | 56–85 | 4–17 | 9–18 |
| Dhanipur | 38–76 | 11–61 | 1–41 | 52–84 | 10–45 | 3–18 | 45–75 | 24–54 | 1–25 |
| Geonkhali | 46–78 | 9–40 | 12–38 | 61–87 | 9–18 | 4–19 | 49–74 | 10–32 | 16–21 |

(*Source* Field survey and laboratory experiment)

distributional pattern more chaotic and haphazard (Pettijohn 1975). Wai et al. (2004) mentioned that the grain size distribution is greatly affected by tidal pumping and tidal trapping in estuarine environment.

## 6.5   Identification of Sediment Type by Triangular Diagram

Typically large and complex sedimentological datasets can be easily compiled into tables, but it is very difficult to understand and interpret the pure numerical information of the datasets (Venkatramanan et al. 2010). Thus, most of the scientists prefer to use graphical representations of the datasets to reduce complexities, understand trends and patterns in the data and try to develop hypothesis. Generally, the sedimentologists use the *equilateral triangular diagrams* as a common graphical technique to represent the percentages of gravel, sand, silt, and clay present in a sediment sample (Venkatramanan et al. 2010). This facilitates the presentation of sedimentological data in a simple and understandable way and becomes helpful for the easy classification of sediments and comparison of sediment samples (Poppe and Eliason 2007). Folk (1974) also attempted to classify the sediment samples by plotting the percentage of sand, silt and clay in a triangular diagram. In the area under study 81.33% of the sediment samples are of silty sand type, 7.33% are of muddy sand, 6% samples fall in sandy silt category and remaining 4.66% of sediment samples are other type (Fig. 6.7). In monsoon season, 90% of sediment samples are of silty sand type, while in pre-monsoon and post-monsoon season only 75% of the samples are of silty sand type (Fig. 6.7).

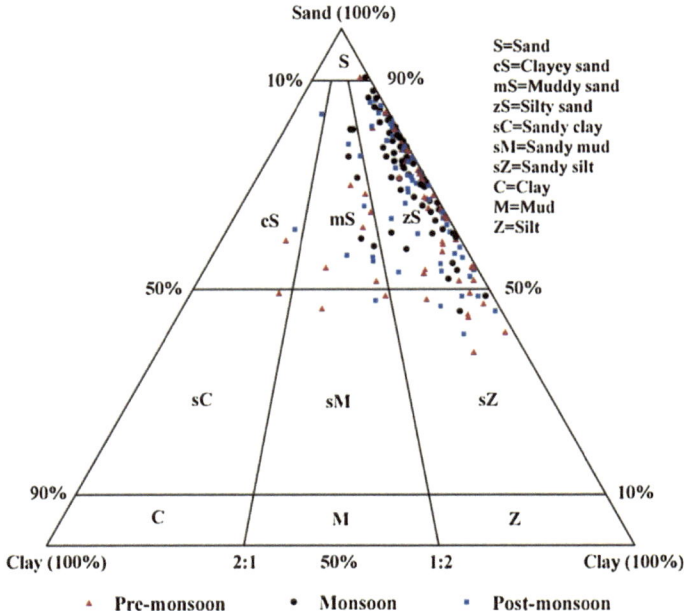

**Fig. 6.7** Ternary diagram showing the nature of sediments (Folk 1980). (*Source* Field survey and laboratory experiment)

# References

Amaral EJ, Prayor WW (1977) Depositional environment of the St. Peter sandstone deduced by textural analysis. J Sediment Petrol 47:32–52

Asselman NEM (1999) Grain size trends used to assess the effective discharge for flood plain sedimentation, River Waal, The Netherlands. J Sed Res 69:51–61

Awasthi AK (1970) Skewness as an environmental indicator in the Solani River system, Roorkee, India. Sediment Geol 4:177–183

Baruch J, Kotoky P, Sarma JN (1997) Textural and geochemical study on river sediments: a case study on the Jhanji river, Assam. J Indian Assoc Sedimentologists 16:195–206

Bhatia MR, Cook KAW (1986) Trace element characteristics of graywackes and tectonic setting discrimination of sedimentary basins. Contrib Mineral Petrol 92:181–193

Brambati A (1969) Stratigraphy and sedimentation of Siwaliks of North Eastern India. Proc Inter Sem Intermontane Basins: Geology and Resources, Chiang Mai, Thailand, pp 427–439

Buffington J, Montgomery D (1997) A systematic analysis of eight decades of incipient motion studies, with special reference to gravel—bedded rivers. Water Resour Res 33(8):1993–2029

Church M (2006) Bed material transport and the morphology of alluvial river channels. Annu Rev Earth Planet Sci 34:325–354

Cronan DS (1972) Skewness and Kurtosis in polymodal sediments from the Irish Sea. J Sediment Petrol 42(1):102–106

Dietrich WE, Kirchner JW, Ikeda H et al (1989) Sediment supply and the development of the coarse surface layer in gravel-bedded rivers. Nature 340(6230):215–217. doi:10.1038/340215a0

Duane DB (1964) Significance of skewness in recent sediments, western Pamlico Sound, North Carolina. J Sediment Petrol 34(4):864–874

Folk RL (1974) Petrology of sedimentary rocks. Hemphill Publishing Co, Austin

Folk RL (1980) Petrology of sedimentary rocks. Hemphill Publishing Co, Austin

Folk RL, Ward MC (1957) Brazos River bar (Texas): a study in the significance of grain size parameters. J Sediment Petrol 27(1):3–27

Fralick PW, Kronberg BI (1997) Geochemical discrimination of clastic sedimentary rock sources. Sediment Geol 113:111–124

Friedman GM (1961) Distinction between dune, beach and river sands from their textural characteristics. J Sediment Petrol 31(4):514–529

Friedman GM (1962) On sorting, sorting co -efficient and log—normality of the grain size distribution of sandstones. J Geol 70:737–753

Friedman GM (1967) Dynamic processes and statistical parameters compared for size frequency distribution of beach river sands. J Sediment Petrol 37(2):327–354

Ghosh SK, Chatterjee BK (1994) Depositional mechanism as revealed from grain size measures of the Palaeoproterozoic Kolhan Siliciclastics. Keonjhar District, Orissa, India. Sediment Geol 89:181–196

Inman DL (1952) Measures for describing the size distribution of sediments. J Sediment Petrol 22 (3):125–145

Maity SK (2015) Cognition of interworking of processes leading to sedimentation at lower reach of the Rupnarayan River, West Bengal, India. Dissertation, Vidyasagar University, West Bengal, India

Martins LR (1965) Significance of Skewness and Kurtosis in environmental interpretation. J Sediment Petrol 35(3):768–770

Martins LR (1967) Aspectos texturais e deposicionais dos sedimentos praiais e eolicos da Planicie Costeira do Rio Grande do Sul. Escola de Geologia. UFRGS. Publicacao Especial 13:102

Martins LR (2003) Recent sediments and grain-size analysis. Gravel 1:90–105

Mason CC, Folk RL (1958) Differentiation of beach, dune and aeolian flat environments by size analysis, Mustang Island, Texas. J Sediment Petrol 28:211–226

McLaren P, Bowels SD (1985) The effects of sediment transport on grain size distribution. J Sediment Petrol 55(4):457–470

Muraleedharan Nair MN, Ramachandran KK (2002) Textural and trace metal distribution in sediments of Beypore estuary. Indian J Mari Sci 31:295–304

Pettijohn FJ (1975) Sedimentary rocks, 3rd edn. Harper & Row, New York

Poppe LJ, Eliason AH (2007) A Visual Basic program to plot sediment grainsize data on ternary diagrams. Comput Geosci 34:561–565

Rajgnapathi VC, Jitheshkumar N, Sundararajan M, Bhat KH, Velusamy S (2012) Grain size analysis and characterization of sedimentary environment along Thiruchendur coast, Tamilnadu, India. Arab J Geosci 23(3):45–56

Rashi M, Vetha Rao D, Chandrasekhar N (2011) Tsunami-sediment signatures in the Manakudy Estuary along the West Coast of India. Sci Tsunami Hazards 30(2):94–107

Rhodes RF (1950) Effects of salinity on current velocities. US Corps of Engineers, Committees Tidal Hydraulics, report No-1, p 94

Sahu BK (1964) Depositional mechanism from the size analysis of clastic sediments. J Sediment Petrol 34(1):73–83

Seralathan P, Padmalal D (1994) Textural studies of the surficial sediments of Muvattupuzha river and central Vembanad Estuary, Kerala. J Geol Soc India 43:179–190

Sly PG, Thomas RL, Pelletier BR (1982) Comparison of sediment energy-Texture relationships in marine and lacustrine environments. Hydrobiologia 91:71–84

Valia HS, Cameron B (1977) Skewness as paleoenvironmental indicators. J Sediment Petrol 4:784–793

Venkatramanan S, Ramkumar T, Anitha Mary I (2010) Textural characteristics and organic matter distribution patterns in Tirumalairajanar River Estuary, Tamilnadu, East coast of India. Int J Geomatics and Geosciences 1(3):552–562

Visher GS (1969) Grain size distributions and depositional processes. J Sediment Petrol 39 (3):1074–1106

Wai OWH, Wang CH, Li YS et al (2004) The formation mechanisms of turbidity maximum in the Pearl River estuary. China Mar Pollut Bull 48:441–448

Wang et al (1998) Cross shore distinction of sediment texture under breaking waves along low energy coasts. J Sediment Res 68:497–506

# Chapter 7
# Conclusion

Presently the *lower reach* of the *Rupnarayan River*, from Kolaghat to Geonkhali (40 km) has been showing signs of rapid deterioration and incapacitation due to *sedimentation* that creates a series of inter-connected problems like shifting of river course, shortage of water resources, hindrance of easy discharge of water and resultant flood, navigation difficulties, river bank erosion and loss of settlement and properties. Detailed studies on *causes, mechanisms* and *magnitude* of sedimentation reveal that the *asymmetry* of the cross-sections leads to channel dynamism due to concentration of energy near the bank. *Channel widening* and *separation of flow* and associated variation in energy distribution across the channel lead to sedimentation. Sudden widening of the channel near Kolaghat leads to flow separation, reduction of energy and deposition of sediment (during low tide) and near Geonkhali region, sudden constriction (bottle neck shape) of the channel (width-depth ratio—130.43) hinders the free draining of ebb tide water leading to ponding effect and velocity reduction that leads to sudden reduction of energy and invites sedimentation. Seasonal variation of *water discharge* is very important to control the stream energy, sediment transporting capacity and rate of sedimentation. During dry season paucity of rainfall causes less discharge of water (850–4160 m$^3$/s), reduction of stream energy and sediment transporting capacity which allows sedimentation. But in monsoon season occurrence of huge rainfall increases the water discharge (3455–9050 m$^3$/s), stream energy and sediment transporting capacity. Because of this sedimentation rate is high during dry season but less in monsoon season. Again, high tide duration is shorter by 2–6 h than that of low tide and this *tidal asymmetry* results swifter flow with greater energy during high tide leading to landward transport of sediment. The sluggish low tide discharge over longer duration (8–9 h) allows sufficient opportunity for settlement of sediments.

During non-monsoon period, upstream penetration of suspended sediment is more during high tide than that is discharged towards downstream during low tide which accelerates the rate of *sedimentation*. But in monsoon, the transport of suspended sediment during high tide and low tide is almost equal which restricts the *sedimentation* rate. Increasing tendency of water velocity and discharge towards

© The Author(s) 2018
S. Kumar Maity and R. Maiti, *Sedimentation in the Rupnarayan River*,
SpringerBriefs in Earth Sciences, https://doi.org/10.1007/978-3-319-62304-7_7

downstream leads to the increase of bed load transport rate from Kolaghat to Geonkhali. In non-monsoon season, the transport of bed load is more during high tide than during low tide, but the transport of bed load is almost equal in both the tidal phases in monsoon season. Sediments are coarser in monsoon than in pre-monsoon and post-monsoon seasons due to increase of water volume, stream energy and removal of fine sediments in monsoon. In dry season, sediments are mostly moderately to well sorted but in monsoon season sediments are mostly poorly to very poorly sorted. Generally, the coarser sediments are negatively skewed and finer sediments are positively skewed. So, the mechanisms of sedimentation, the rate of sediment transport and textural characteristics of sediments in the lower reach of the Rupnarayan River are the result of the interactions between fluvial and marine processes and the seasonal variation of the intensity of their influences.

The findings and outcomes of the work will be supportive and encouraging to the engineers, hydrologists, planners and other concerned authorities to take rational decision for the controlling of sedimentation rate and *management of associated problems* not only in the study area but also in any of the tidal river in the world.

# Index

© The Author(s) 2018
S. Kumar Maity and R. Maiti, *Sedimentation in the Rupnarayan River*,
SpringerBriefs in Earth Sciences, https://doi.org/10.1007/978-3-319-62304-7